# Heat Stress In Food Grain Crops: Plant Breeding and Omics Research

## Edited by

### Uday C. Jha

*CI Division, Indian Institute of Pulses Research,*
*Kanpur,*
*Utar Pradesh,*
*India*

### Harsh Nayyar

*Department of Botany,*
*Panjab University,*
*Chandigarh,*
*India*

### &

### Sanjeev Gupta

*CI division,*
*Indian Institute of Pulses Research,*
*Uttar Pradesh*
*India*

# Heat Stress In Food Grain Crops: Plant Breeding and Omics Research

Editors: Uday Chand Jha, Harsh Nayyar and Sanjeev Gupta

ISBN (Online):  978-981-14-7398-2

ISBN (Print):  978-981-14-7396-8

ISBN (Paperback): 978-981-14-7397-5

need for a court order if at any point you breach any terms of this License Agreement. In no event will any delay or failure by Bentham Science Publishers in enforcing your compliance with this License Agreement constitute a waiver of any of its rights.

3. You acknowledge that you have read this License Agreement, and agree to be bound by its terms and conditions. To the extent that any other terms and conditions presented on any website of Bentham Science Publishers conflict with, or are inconsistent with, the terms and conditions set out in this License Agreement, you acknowledge that the terms and conditions set out in this License Agreement shall prevail.

**Bentham Science Publishers Pte. Ltd.**
80 Robinson Road #02-00
Singapore 068898
Singapore
Email: subscriptions@benthamscience.net

**BENTHAM SCIENCE**

# CONTENTS

# PREFACE

High temperature during the growth period of most crop plants causes negative impact from crop germination, vegetative and reproductive stage to grain development, thus causing serious crop yield penalty. Exposure to high temperature can change different metabolic functions. Research on heat sensitive cultivars showed that heat stress for longer duration inhibits Rubisco activity and reproductive functions and thus produces an array of changes in plants. The insights on various mechanisms in plants for adaptation is crucial for developing resilience to high temperature.

The present book is an excellent review of recent advances in research on analyzing negative impacts of heat stress challenging crop yield and intervention of various studies to overcome the challenge of heat stress. To overcome the challenges of heat stress, the authors have elaborated on the various approaches, including conventional plant breeding, physiological trait-based breeding approach, various 'omics' based approaches covering genomics, transcriptomics, proteomics, metabolomics and ionomics. Efforts have been made to highlight the scope of emerging novel breeding schemes *viz.*, genomic selection, and genome editing tools for improving genetic gain in crop plant.

The book contains chapters authored by scientists/researchers who are actively involved in improving the yield of agricultural crops by mitigating heat stress. Their contribution is enormous in presenting up-to-date information on the subject. The book will be beneficial to plant breeders, molecular biologist and plant physiologist as it gives insights into advanced breeding schemes, discovery of novel candidate gene(s)/QTLs related to heat stress tolerance and various adaptive mechanisms working at physiological and biochemical level mediating heat tolerance in plant. Thus, the information contained in this book will enrich our understanding of various pathways, genes rendering heat tolerance in plants and also helps us to develop various strategies to ensure global food security against heat stress.

We editors are thankful to our parent organization, Indian Council of Agricultural Research (ICAR), New Delhi for supporting our scientific pursuit in the form of a book "Heat Stress In Food Grain Crops: Plant Breeding and Omics Research". We are highly thankful to Dr. T. Mohapatra, Director General, ICAR and Secretary, DARE, Ministry of Agriculture and Farmers' Welfare, Government of India and Dr. T.R. Sharma, Deputy Director General (Crop Science), ICAR for their constant support and encouragement in this endeavor.

We thank our families for being patient and supportive in this long journey, without their moral support, it would not be possible. The entire team at Bentham, especially the Publishing Editor, and Production Editor who have always been cooperative to make this publication a reality. They have been very generous in accommodating even last minutes changes and deserve our genuine appreciation. We hope that this book will absolutely serve its purpose and will provide a latest and comprehensive treatise to the readers in furthering their academic and research pursuits.

Kanpur, the 16th August 2020

**Uday C. Jha**
CI Division, Indian Institute of Pulses Research,
Kanpur
Uttar Pradesh
India

# LIST OF CONTRIBUTORS

| | |
|---|---|
| **Aditya Pratap** | Indian Institute of Pulses Research, Kanpur, India |
| **Akshita Tyagi** | Division of Genetics, ICAR-Indian Agricultural Research Institute, New Delhi, India |
| **A. G. Vijaykumar** | Seed Unit, UAS Dharwad, India |
| **B. S. Patil** | Regional Centre, ICAR-Indian Agricultural Research Institute, Dharwad, India |
| **C. Bharadwaj** | Division of Genetics, IARI, New Delhi, India |
| **Chikkappa G. Karjagi** | Unit Office, ICAR-Indian Institute of Maize Research, New Delhi, India |
| **Debjoti Sengupta** | Indian Institute of Pulses Research, Kanpur, India |
| **Dnyaneshwar B. Deshmukh** | International Crops Research Institute for the Semi-Arid Tropics (ICRISAT), Hyderabad, India |
| **Firoz Hossain** | Division of Genetics, ICAR-Indian Agricultural Research Institute, New Delhi, India |
| **Gyanendra Singh** | Indian Institute of Wheat and Barley Research, India |
| **Gyanendra Pratap Singh** | Indian Institute of Wheat and Barley Research, India |
| **Harsh Nayyar** | Department of Botany, Panjab University, Chandigarh, India |
| **H. C. Lohithaswa** | Department of Genetics and Plant Breeding, College of Agriculture, (University of Agricultural Sciences, Bengaluru, India |
| **Janila Pasupuleti** | International Crops Research Institute for the Semi-Arid Tropics (ICRISAT), Hyderabad, India |
| **Jayant S. Bhat** | Regional Centre, ICAR-Indian Agricultural Research Institute, Dharwad, India |
| **Kadambot H.M. Siddique** | The UWA Institute of Agriculture, M082, Perth, Australia |
| **Karnam Venkatesh** | ICAR, Indian Institute of Wheat and Barley Research, India |
| **M.G. Mallikarjuna** | Division of Genetics, ICAR-Indian Agricultural Research Institute, New Delhi-110012, India |
| **Mamrutha H.M.** | ICAR-Indian Institute of Wheat and Barley Research, India |
| **Manu Priya** | Department of Botany, Panjab University, Chandigarh, India |
| **Murali T. Variath** | International Crops Research Institute for the Semi-Arid Tropics (ICRISAT), Hyderabad, India |
| **Narendra Pratap Singh** | Indian Institute of Pulses Research, Kanpur, India |
| **Palanisamy Veeraya** | Division of Genetics, ICAR-Indian Agricultural Research Institute, New Delhi, India |
| **P. Sanjana Reddy** | ICAR-Indian Institute of Millets Research, Hyderabad, India |
| **Seltene Abady** | International Crops Research Institute for the Semi-Arid Tropics (ICRISAT), Hyderabad, India<br>Haramaya University, Addis Ababa, Ethiopia |

| | |
|---|---|
| **Senthilkumar K.M.** | ICAR, Central Tuber Crops Research Institute, India |
| **Sunil Chaudhari** | International Crops Research Institute for the Semi-Arid Tropics (ICRISAT), Hyderabad, India |
| **Swathi Gattu** | International Crops Research Institute for the Semi-Arid Tropics (ICRISAT), Hyderabad, India |
| **Uday Chadn Jha** | Indian Institute of Pulses Research, Kanpur, India |
| **Vikas Gupta** | ICAR, Indian Institute of Wheat and Barley Research, India |

**CHAPTER 1**

# Mitigating Heat Stress in Wheat: Integrating Omics Tools With Plant Breeding

**Karnam Venkatesh[1,*], Vikas Gupta[1], Senthilkumar K.M.[2], Mamrutha H.M.[1], Gyanendra Singh[1] and Gyanendra Pratap Singh[1]**

[1] *ICAR - Indian Institute of Wheat and Barley Research, India*

[2] *ICAR - Central Tuber Crops Research Institute, India*

**Abstract:** Wheat crop is adapted to cooler climatic conditions and has an optimal daytime growing temperature of 15 °C during the reproductive stage. Heat stress is becoming a major constraint to wheat production as it affects every stage of the crop but the anthesis and reproductive stages are more sensitive. The situation will be aggravated due to climate change as predicted by the Intergovernmental Panel on Climate Change, for every degree rise in temperature above this optimum leads to a 6% yield reduction. Being quantitative in nature, heat stress is a complex trait and is strongly influenced by genotype x environment interaction. The new omics approaches like transcriptomics, proteomics, metabolomics and ionomics will be useful in understanding the underlying mechanism of heat tolerance. In this chapter, we will summarize the impact of heat stress on wheat production, physiological traits contributing to heat tolerance and how to integrate new omics tools such as transcriptomics, proteomics, metabolomics and ionomics with plant breeding.

**Keywords:** Chromosome substitution lines, Conventional plant breeding, Multipronged approach, Osmoprotectant molecules, Temperature stress, Transcriptomics.

## INTRODUCTION

Wheat being cultivated as a major staple crop from the prehistoric times, caters to the energy requirement of the human population in India and across the globe (Sharma *et al.* 2015). Wheat improvement efforts in the form of conventional breeding aimed at yield enhancement in the past have led to significant growth in productivity and production.

---

\* **Corresponding author Karnam Venkatesh:** ICAR - Indian Institute of Wheat and Barley Research, India; E-mail:karnam.venkatesh@icar.gov.in

**Uday C. Jha, Harsh Nayyar and Sanjeev Gupta (Eds.)**

However, there is an increased demand for wheat due to changes in consumption patterns in the form of increased demand for wheat based end products such as biscuits, noodles, pasta, *etc.* According to FAO estimates, globally around 840 million tons of wheat must be produced by 2050 from the current levels. Further climate change scenarios in the form of increased heat and drought stress events would pose serious constraints for the achievement of 2050 targets (Reynolds *et al.* 2009). The global climate change in the form of elevated $CO_2$ concentration, warming temperatures, and changes in rainfall patterns is becoming a major threat to crop production (IPCC 2007). The increased events of temperature rise in both the ocean and on the earth until 2012 has been reported (Team *et al.* 2014). The severe and more abrupt rise in temperatures in several parts of the world led to severely reduced crop yields (Kaushal *et al.* 2016). The adverse effect of increased temperatures on plant growth mechanisms is higher especially in the arid and semi-arid regions of the world (Cooper *et al.* 2009). The vulnerability of especially heading and grain filling period in wheat to high temperature stress has been reported (Liu *et al.* 2016; Yang *et al.* 2017; Priya *et al.* 2018).

The emergence and development of automated sequencing methods started the era of omics in the form of genomics and led to the sequencing of the whole genome of *Arabidopsis thaliana* in 2000 (Initiative and others 2000). Later on, several other organisms and crop plant genomes such as rice (Goff *et al.* 2002), soybean (Schmutz *et al.* 2010), maize (Schnable *et al.* 2009) and even the most complicated polyploidy species such as wheat (Consortium and others 2018) were sequenced and made the latest omics tools amenable to crop improvement. The word "omics" formally refers to a study related to genome, proteome, or metabolome, and aims at the characterization of a large family of cellular molecules and exploring their roles, and their interactive effects in an organism. These omics approaches are mainly performed through the application of several high-throughput technologies that mainly involve qualitative and/or quantitative detection of novel or previously identified genes, transcripts, proteins, and metabolites and other molecular species through genomics, transcriptomics, proteomics, and metabolomics, respectively (Ebeed 2019). Application of various omics approaches in understanding the abiotic stress responses in general (Kole *et al.* 2015; Meena *et al.* 2017; Lamaoui *et al.* 2018; Ebeed 2019; Wani 2019), drought stress (Hasanuzzaman *et al.* 2018; Ding *et al.* 2018) and heat stress (Xu *et al.* 2011; Jacob *et al.* 2017; Salman *et al.* 2019) particularly in crop plants and their mitigation has been reported by earlier researchers. It is therefore suggested that a multi-disciplinary and multi-pronged approach integrating the conventional plant breeding with the latest omics tools will be useful in mitigating the adverse effects of heat stress on wheat production. This chapter briefly deals with the latest reports of the application of omics approaches in improving wheat tolerance for heat stress.

## IMPACT OF HEAT STRESS ON WHEAT PRODUCTION

### Heat Stress, Extent of Damage/Threat to Wheat Area and Mechanisms Affected

The climate predictions by the Intergovernmental Panel on Climate Change (IPCC) indicated that the mean atmospheric temperatures are expected to increase between 1.8 to 5.8°C by the end of this century (IPCC 2007). The increase in the frequency of hot days and greater variability in temperatures in the future is also predicted as an effect of climate change (Pittock *et al.* 2003; Team *et al.* 2014). Extreme temperatures directly influence crop production by specifically affecting plant growth and yield realization posing a serious threat to food production (Team *et al.* 2014). Higher temperatures are likely to affect around seven million hectares of wheat area in developing countries and around 36 million hectares in temperate wheat production countries (Reynolds 2001). Warmer temperatures resulted in an annual wheat yield reduction to the tune of 19 million tons amounting to a monetary loss of $2.6 billion was observed between 1981-2002 (Lobell and Field 2007). In India, it has been predicted that with every rise in 1°C temperature, the wheat production will be decreased by 4–6 million tonnes (Ramadas *et al.* 2019). Approximately, 3 million ha wheat area in northeastern and northwest plain zones is exposed to terminal/reproductive heat stress (Gupta *et al.* 2013). Another report by Joshi *et al.*, (2007) stated that around 13.5 million ha wheat area in India is vulnerable to heat stress. Temperatures above 34°C in northern Indian plains leading to significant yield loss was reported (Lobell *et al.* 2012).

High temperature stress when occurred at germination and early establishment stages is known to decrease germination and seedling emergence leading to abnormal seedlings, poor vigour, reduced overall growth of developing seedlings (Essemine *et al.*, 2010; Kumar *et al.*, 2011; Piramila *et al.*, 2012). Further high temperature stress is found to severely impact dry matter partitioning, reproductive organ development and reproductive processes in crop plants (Prasad *et al.* 2011). Intermittent spells of temperature above 30 °C during the reproductive stage causes high temperature stress leading to decreased seed set and low grain number (Prasad and Djanaguiraman 2014; Sehgal *et al.* 2018; Qaseem *et al.* 2019). There are reports which also indicate deterioration of grain quality parameters under high temperature stress (Britz *et al.* 2007).

Grain filling is an essential growth stage involving mobilization and transport processes involving many biochemical processes regulating the synthesis of proteins, carbohydrates and lipids and their transport into the developing grains (Awasthi *et al.* 2014; Farooq *et al.* 2017). Processes leading to grain filling and the accumulation of reserves in the developing grains are highly sensitive to temperature changes (Yang and Zhang 2006). Heat stress affects enzymatic processes involved in the synthesis of starch and proteins and ultimately affecting the transport and accumulation of major components of grains primarily the starch and proteins (Asthir *et al.* 2012; Farooq *et al.* 2017).

Heat stress in wheat can lead to early senescence thereby reducing the time available for grain filling (Awasthi *et al.* 2014). The senescence effect is accelerated by heat disrupting chloroplasts and damaging chlorophyll and the leaf membranes and increase in ethylene production (Prasad and Djanaguiraman 2014) which further reduces photosynthetic efficiency, biomass accumulation and yield attainment. Under normal conditions, the photosynthetic assimilates accumulated during the pre-anthesis period in the form of stem reserves contribute to around 10-40% of final grain weight. (Gebbing and Schnyder 1999). Remobilization of these stem reserves to the grains is crucial to attain full grain size and yield (Asseng and van Herwaarden 2003). Heat stress led accelerated canopy senescence reduces photosynthetic area and hence source strength clubbed with reduced turgor in phloem cells due to water deficiency, thereby increasing the viscosity of sucrose inhibiting its transport through phloem toward the grains (sink)(Sevanto 2014). Heat causing a reduction in activities of PEP carboxylase and RuBP carboxylase leading to inhibition of carbon assimilation in maturing grains due to heat stress was also observed in wheat (Xu *et al.* 2004). Heat stress during grain filling markedly decreasing starch accumulation in wheat by altering the expression of starch-related genes leading to a reduction in seed size in wheat (Hurkman *et al.* 2003; Dupont and Altenbach 2003).

## MECHANISM OF HEAT TOLERANCE ADOPTION BY PLANTS

### Avoidance Mechanisms by Way of Phenotypic Adjustments

Crop plants as part of their adaptation mechanisms to higher temperatures display a greater level of phenotypic plasticity. Plants adapt to higher temperatures by way of certain morphological adjustments in its life cycle, increased pubescence (Maes *et al.* 2001; Banowetz *et al.* 2008), increased wax deposition of leaf, sheath and on the stem surface, changed leaf orientation, manipulation of membrane lipid fractions, *etc.* First and the foremost adaptation of the plant when heat is sensed is shortening its life cycle to escape the adverse effect of stress (Blum *et al.*, 2001). Leaf rolling to reduce the excess loss of water through transpiration is also one of the adaptation mechanisms found in wheat (Sarieva *et al.* 2010). Further as the reproductive growth stages of wheat (flowering and grain development) are the most sensitive stages to heat stress and the plant is forced to quickly complete these stages thereby shortening the whole life cycle (Hall 1992; Hall 1993). Water conservation mechanisms such as increased wax deposition on the plant surfaces has been observed under high temperature conditions and is linked to several favourable effects on plant in the form of protection against excess radiation and also contributes to reflection of visible and infrared wavelengths of light thereby reducing evaporative water loss through plant surfaces (Shepherd and Wynne Griffiths 2006; Cossani and Reynolds 2012). The wax is also known to reduce the leaf temperatures thereby protecting the membranes and leaf structural components from heat damage (Mondal 2011). Larger xylem vessels enabling plants to compensate for increased water loss under high temperature is also an adaptive mechanism (Srivastava *et al.* 2012). Under well-watered conditions increased transpiration leading to reduced canopy temperature up to 10°C lower than ambient temperature was an adaptive ability.

High temperature stress effect can also be minimised by manipulating agronomic practices such as using proper sowing methods, proper seed rate, selection of suitable cultivar, increased irrigation frequency, crop mulching *etc.* (Meena *et al.* 2015; Meena *et al.* 2019a; Meena *et al.* 2019b). Examples of managing high temperature stress in wheat by deliberately choosing heat tolerant cultivars (Glennson 81) over heat sensitive (Pavon 76) resulted in higher yields under stress (Badaruddin *et al.* 1999). They further demonstrated that application of animal manure, straw mulch along with increased doses of inorganic nutrients and irrigation frequency achieved enhanced wheat yields under high temperature stress. Foliar spray of potassium orthophosphate ($KH_2PO_4$), calcium, Mg and Zn were reported to enhance high temperature tolerance of wheat (Dias and Lidon 2010; Waraich *et al.* 2011). Practice of integrated approach combining above options could help in minimizing high temperature effects.

**Tolerance Mechanisms**

Ability of the plant to achieve normal growth and produce economic yield under higher temperatures is called as heat tolerance. The plants have evolved various tolerance mechanisms such as altering ion transporter systems, production of late embryogenesis abundant (LEA) proteins, accumulation of osmoprotectant molecules, ion transporters, free-radical scavengers and manipulating systems involving factors like ubiquitin and dehydrin through signaling cascades and transcriptional control (Wang *et al.* 2004; Rodríguez *et al.* 2005). Stomatal closure leading to reduced evaporative water loss to sustain the water dependent plant processes under heat stress (Woodward *et al.* 2002).

Enhanced root development under abiotic stress conditions to reach the deeper layers of soil in order to absorb more water has been reported as an tolerance mechanism (Lehman and Engelke 1993). Extended grain filling duration was also observed as a tolerance mechanism and positive association of grain filling duration with higher yield under heat stress has been reported (Yang *et al.* 2002).

Adjustments in the photosynthetic mechanisms and the enzymes involved by the plants have been found to be an alternate tolerance mechanism adapted by plants. Increased affinity of Rubisco the main enzyme responsible for fixation of carbon to $CO_2$ under higher temperature conditions has been reported in some plants like *Limonium gibertii* (Parry *et al.* 2010). At very high temperatures above optimal, higher activity in the photosynthetic apparatus (Ristic *et al.* 2007; Allakhverdiev *et al.* 2008) and higher carbon allocation and nitrogen uptake rates were seen as tolerance mechanisms (Xu *et al.* 2006).

## TRAITS OF IMPORTANCE FOR HEAT TOLERANCE AND THEIR PHENOTYPING TECHNIQUES

### Canopy Temperature

Reduction in the temperature of the crop canopy under high temperature stress as a result of increased transpiration has been found to be an important trait to be associated with heat tolerance (Cossani and Reynolds 2012). The ability of the plant to maintain a cooler canopy was found to be genetically controlled and therefore amenable for selection of germplasm lines with cooler canopies (Pinto *et al.* 2010). Canopy temperature (CT) can be easily measured for germplasm screening using an infrared thermometer. The infrared thermometer senses this radiation and converts into electrical signal and is displayed as temperature. CT measurement by infrared thermometer being a non-destructive method can be used under field conditions and can covers large number of genotypes and selection for this trait indirectly allows for the selection of genotypes with better water use, deep root and stomatal conductance under stress.

### Leaf Chlorophyll Content

The crop canopy greenness contributed mainly by the photosynthetic pigment chlorophyll is another trait of importance to screen germplasm for heat tolerance. The chlorophyll pigment reflects only the green fraction of the light after absorbing all other colour fractions and hence it is green in colour. The canopy greenness is directly related to photosynthetic efficiency of the plants. The chlorophyll content of the leaf can be estimated by a destructive lab based DMSO: acetone extraction method and by using an instrument called chlorophyll meter which is non-destructive and optical method. The measurement by optical method using different types of chlorophyll meters is found to be more relevant than DMSO method under field conditions (Dwyer *et al.* 1991). The chlorophyll content measured through chlorophyll meters is in the form of an index called chlorophyll content index (CCI). The CCI ranges from 0 to 99.9 and with the increase in the level of heat stress the CCI decreases and CCI of healthy plant ranges from 40 to 60. As optical method is based on leaf reflectance, it is influenced by time of day in terms of light (Mamrutha *et al.* 2017). Care should be taken to measure chlorophyll content at uniform time and in specific leaf across the genotypes under field (Mamrutha *et al.* 2017).

## Canopy Greenness/Stay Green Canopy

Prolonged maintenance of canopy greenness also referred to as stay-green nature is a physiological adaptation mechanism by plants under heat stress. Lim *et al.*, (2007) described stay greenness as "leaf senescence is characterized initially by structural changes in the chloroplast, followed by a controlled vacuolar collapse, and a final loss of integrity of plasma membrane and disruption of cellular homeostasis". Stay green trait in tolerant genotypes help in withstanding chlorophyll loss and maintain photosynthesis levels under high temperature stress. Association of stay green habit with sustained yield levels under heat stress has been earlier reported and QTL regions regulating this have been identified (Vijayalakshmi *et al.* 2010). There are mainly two types of stay green types. One is productive type, where in the stay green plant parts actually contribute for sink/ grain filling. Another is cosmetic stay green type, where in greenness in these plants will not contribute for grain filling. Hence, identification of true and productive stay green types are also a challenge and can be done by considering other traits like water soluble carbohydrates in stem, peduncle *etc.* (Mamrutha *et al.*, 2019).

The canopy greenness can be measured by an instrument known Normalized difference vegetation index (NDVI) sensor. Spectral reflectance based NDVI values (range between 0 to 1) are highly correlated with yield under temperature stress (Lopes and Reynolds 2012). Zero represents no greenness and one represents maximum greenness (Mamrutha *et al.* 2017). stay green habit can also be measured by other instruments such as canopy analyser (Licor) or porometer which measures leaf area index and green area index (GAI). Many other techniques like the digital photography of the canopy can also be taken from same height from the ground level and pictures can be analysed with different softwares (Adobe photoshop CS3 extended or later version) to assess the early ground cover (Mullan and Reynolds 2010).

## Earliness *Per se* in Wheat

Earliness (earliness *per se*) in wheat is an adaptation strategy characterized by early heading followed by early maturity of genotypes under high temperature stress environments. Earliness helps genotypes to complete the essential plant growth stages such as seed setting and grain filling under favourable temperatures thereby avoiding the occurrence of terminal/late heat stress. Mondal *et al.* (2013) reported that the early heading entries performed well in areas affected from terminal heat stress as earliness helps them to escape high temperatures during grain filling stages. In addition to helping them escape the terminal heat stress, earliness also resulted in achieving >10% higher yield compared to the local check varieties under high temperature stress environments. High grain filling rate in early maturing gentotypes was also reported to be promoting heat stress tolerance in durum wheat (Al-Karaki 2012). Tewolde *et al.* (2006) reported that earliness helped cultivars adopt to high temperature stress as they had longer post-heading period resulting in longer grain filling duration. Therefore, earliness was also suggested as a key trait in breeding for high temperature stress tolerance (Joshi *et al.* 2007b).

**Photosynthetic Efficiency**

The differential rate of photosynthesis expressed as photosynthetic efficiency is again a very essential component trait contributing to tolerance under high temperature stress. Stable photosynthetic rates over longer duration in heat tolerant genotypes contributed to higher grain weight, higher harvest index under stress showing the positive association of rate of photosynthesis with yield parameters under heat (Al-Khatib and Paulsen 1990). Looking at the major role played by photosynthesis in determining yield under heat stress, it is also pertinent to have phenotyping techniques to help breeders to select for genotypes with higher photosynthetic efficiency. The relative photosynthetic efficiency can be indirectly predicted using the chlorophyll content index, however there are instruments available which can measure the photosynthesis exactly. Infra-red gas analyser (IRGA) is used to measure the photosynthesis on a real time basis when stress period is available or stress is imposed under experimental conditions. IRGA measures the amount of $CO_2$ fixed during photosynthesis by estimating the difference in amount of $CO_2$ pumped in and moving out of closed leaf chamber (Nataraja and Jacob 1999). Photosynthesis using IRGA should be measured at noon or prior or after noon to get maximum photosynthesis and to avoid error and it should be recorded at uniform positions in the leaf.

**Chlorophyll Fluorescence (CFL)**, is also one of the traits used extensively to indirectly measure the photosynthetic efficiency of the genotypes. It is used, mainly as indicator of Photo system II (PSII) function. Applicability of CFL in screening wheat genotypes for heat tolerance and its use in accumulating genes favouring heat tolerance is well-known (Moffatt *et al.* 1990; Dash and Mohanty 2001). CFL meters are used to measure the CFL and they measure Fv/Fm ratio *i.e.* immediately after dark adaptation when leaf is exposed to light. The maximum amount of photons used for photochemistry is estimated as ratio of Fv/Fm where in Fv-Variable fluorescence and Fm-Maximal fluorescence. When photons fall on the leaf surface, it is being dissipated mainly into two processes *i.e.* photochemical quenching in the form of photosynthesis and non-photochemical quenching in the form of heat and fluorescence. When the plant is stressed, the PSII efficiency will be reduced and hence will get less value of the ratio compared to tolerant genotypes (Maxwell and Johnson 2000).

## Cell Membrane Thermal Stability

Under high temperature conditions the cell membrane becomes weak and tends to rupture leading to leakage of electrolytes. Membrane thermal stability is being repeatedly used as a measure of electrolyte diffusion resulting from heat induced cell membrane leakage. Increased level of electrolyte leachates diffused from cells is measured here. Heat tolerant genotypes are identified by measuring electrical conductivity as an index to indirectly measure membrane thermal stability (Blum and Ebercon 1981; Saadalla *et al.* 1990; Blum 2018). Greater amount of electrical conductivity said to be indicating better heat-stress tolerance (Saadalla *et al.* 1990). Presence of high genetic heritability of membrane stability in wheat was seen to be an advantage for its use in breeding for heat tolerance (Fokar *et al.* 1998; Reynolds 2001). An electrical conductivity meter can be used to measure the membrane thermal stability with a reference standardizing solution of 0.005N KCl (Ibrahim and Quick 2001; Bala and Sikder 2017; ElBasyoni *et al.* 2017).

## WHEAT IMPROVEMENT FOR HEAT TOLERANCE, INTEGRATING PLANT BREEDING AND OMICS TOOLS

### Breeding for Heat Tolerance in Wheat

Heat tolerance of wheat can be improved by following conventional plant breeding approaches like identification of superior germplasm followed by hybridization and pedigree selection. Molecular marker assisted selection wherein conventional breeding clubbed with indirect selection of superior genotypes using linked DNA-based markers.

### Germplasm Identification for Heat Tolerance

The differential performance of genotypes for yield traits under heat stress has been widely used as a selection criterion to identify tolerant wheat genotypes. The genetic variability available in the wheat gene pool in the form of wild progenitor species and synthetics can be utilized for continued improvement of wheat adaptation to abiotic stress. Sareen *et al.*, (2012) identified four heat tolerant lines which from screening of synthetic wheat derivative germplasm. Wild relatives of wheat also present a rich source of diversity. *Triticum dicoccoides* and *T. monococcum* have been reported as potential sources of germplasm that can be used to enhance heat tolerance in bread wheat. Additionally, variable degrees of heat tolerance were observed in *Aegilops speltoides, Ae. longissima Ae. taushii* and *Ae. Searsii* (Ehdaie and Waines 1992; Waines 1994; Zaharieva *et al.* 2001; Pradhan *et al.* 2012; Awlachew *et al.* 2016). Awlachew *et al.*, (2016) developed backcross introgressed lines (BILs) using heat-tolerant accession of *Ae. speltoides* pau3809 as one of the parent, these BILs showed improvement over heat tolerance component traits compared to recurrent parent. One of the most widely used wheat relatives is rye (*Secale cereale* L.), which is well-documented as a rich source of biotic and abiotic resistance/tolerance (Ehdaie and Waines 1992; Mondal *et al.* 2016).

## Heat Tolerant Varieties by Conventional Selection

Heat tolerance is composed of several component traits and influenced by physiological parameters discussed in the above sections. Presence of greater genotypic variation among the wheat germplasm and involvement of mainly genetic control in the inheritance of these traits gives way for improving heat tolerance *via* plant breeding. The traits regulating yield and yield contributing parameters under high temperature stress are good candidates for targeted breeding for heat tolerance. For example, cooler canopy temperature is shown to be associated with deeper roots as well as higher yield under stress (Pinto *et al.* 2010; Lopes and Reynolds 2012).

The lower canopy temperature was found to be contributed by a cascade of plant processes such as increased stomatal conductance and the presence of genetic variation for this trait is an added advantage (Reynolds *et al.* 2007) and therefore selection for lower canopy temperature can lead to higher yield levels under stress. Additionally, selection of superior parents with better germination (Cargnin *et al.* 2006), early establishment and vigour can also help in developing heat tolerant cultivars (Richards and Lukacs 2002; Mullan and Reynolds 2010). Indian wheat program has released few varieties possessing moderate level of heat tolerance like Raj 3765, UP 2425, DBW16, HW2045, HD2643, DBW 14 and DBW 90 (ICAR-IIWBR 2018). There are a few more popular cultivars adapted to high temperature regions such as C 306 and Lok-1 which are characterised by tall stature and good early vigour with higher shoot and root biomass at seedling stage (Nagarajan and Rane 2000). Recently released cultivars *viz.*, Zakia and Akasha have raised the yield levels in Sudan where the temperatures usually are above 40°C.

## QTLs for Heat Tolerance

Heat tolerance is a quantitative trait, controlled by a number of genes/QTL. Over the last decade's efforts have been made to reveal the genetic basis of heat tolerance. Grain filling duration and yield under heat stress are also important traits to breed for heat tolerance. QTL regions controlling grain filling duration and yield under heat stress have been identified (Pinto *et al.* 2010; Vijayalakshmi *et al.* 2010) which can be used to select heat tolerant progeny in breeding programmes through marker aided selection. Heat susceptibility index (HSI) was used as an indicator of yield stability and QTL controlling HSI were identified (Mason *et al.* 2010) which could be used as a selection criterion in breeding for heat tolerance.

Langdon chromosome substitution lines were firstly used in mapping heat tolerance genes and associated genes were found on chromosomes 3A, 3B, 4A, 4B, and 6A in 1991 (Sun and Quick 1991). Xu *et al.*, (1996) later reported that chromosomes 3A, 3B and 3D were associated with heat tolerance in wheat cultivar (cv) Hope. Using chromosome substitution lines between Chinese Spring and Hope, chromosomes 2A, 3A, 2B, 3B, and 4B of Hope significantly enhanced heat tolerance (Chen *et al.* 2007). Two key QTL on chromosome 3B for canopy temperature and grain yield were detected by Bennett *et al.*, (2012).

## Marker Assisted Recurrent Selection for Improving Heat Stress Tolerance

All the favourable traits governing heat stress tolerance such as early vigour, chlorophyll content, canopy temperature, NDVI, chlorophyll fluorescence, leaf area, grain filling duration, thousand kernels weight, grain yield were accumulated into progeny lines through marker assisted recurrent selection. Following a meticulous inter-mating of $F_5$ progenies carrying different alleles led to accumulation of 4-8 favourable QTLs per progeny and progenies were superior in performance to the parents (Jain *et al.* 2014).

## OMICS APPROACHES FOR IDENTIFICATION OF HIGH TEMPERATURE STRESS TOLERANCE GENES

Recent advances in the next generation sequencing (NGS) based genomics technologies enable to generate huge high-quality genomic data in less time with cost-effective for plant research. Several genes have been identified and functionally validated for high temperature stress tolerance in plants (Yeh *et al.* 2012; Muthusamy *et al.* 2017; Lenka *et al.* 2019). Several groups have employed various omics tools to decipher the molecular mechanism of heat stress tolerance in wheat (Dwivedi *et al.* 2018; Ni *et al.* 2018). Transcriptomics, proteomics, and metabolomics tools were widely used to understand the function of the high-temperature stress responsive genes in wheat (Peng *et al.* 2006; Chauhan *et al.* 2011; Yang *et al.* 2011; de Leonardis *et al.* 2015; Ni *et al.* 2018). Combining omics tools with phenomics would help in understanding the molecular, biochemical and physiological adaptive response of the plants during high temperature stress conditions (Großkinsky *et al.* 2018).

## Transcriptomics

Transcriptome profiling is an extensively used approach to systematically investigate and understand the gene expression during abiotic stress conditions (Katiyar *et al.* 2015; Wang *et al.* 2019). Several groups have deciphered the transcriptome profiles of different tissues of wheat in order to understand the role of genes/QTLs in growth and development (Guan *et al.* 2019). Databases like WheatExp (Pearce *et al.* 2015) and exVIP (Borrill *et al.* 2016; Ramírez-González *et al.* 2018) are developed to study the expression of a gene across different tissues at different developmental stages of wheat. Transcriptome studies revealed the altered expression of ~6000 genes with diverse functions under heat stress in wheat (Qin *et al.* 2008; Vishwakarma *et al.* 2018; Nandha *et al.* 2019; Rangan *et al.* 2020). Heat shock proteins, transcription factors and ribosomal proteins expression were highly altered under heat stress (Vishwakarma *et al.* 2018; Rangan *et al.* 2020). Interestingly, the expression of many genes was modulated by miRNAs during heat stress tolerance, suggesting their role in gene transcription (Ravichandran *et al.* 2019).

## Proteomics

Proteomics approaches harness the robust high throughput technologies for identification and quantification of the proteins in the given tissues (Aslam *et al.* 2017). Proteome analysis helps to decode the structural and functions of different proteins including protein-protein interactions and post-translational modifications (Aslam *et al.* 2017). 309 proteins comprising molecular chaperones (HSP70s, HSP40s, and HSP20s, redox regulatory proteins, proteins involved in metabolic pathways displayed differential expression during heat stress (Wang *et al.* 2018). Zhang *et al.*, (2018) studied the protein profiles of developing wheat kernel under heat stress and identified 78 differentially expressed proteins mainly involved in the signalling and abiotic stress response. These proteins are enriched in the 51 KEGG pathways including protein synthesis, starch and sucrose metabolism. Interestingly, protein profiling of wheat genotypes contrasting in thermotolerance showed the high expression of heat stable proteins and in thermotolerant genotype (Kumar *et al.* 2013). Generation of more high stress responsive proteome profiles from various tissues and development stages would help in deciphering the protein-protein interactions and post-translational modifications in wheat (Kumar *et al.* 2013; Muthusamy *et al.* 2017).

## Metabolomics

High temperature stress severely affects the nutrients including vitamins, minerals, carbohydrates and fats in wheat grains. Thus, studying the metabolite profiles of thermo-tolerant and thermo-sensitive cultivars would help in identification polar molecules including water-soluble organic acids to non-polar lipids (Kumar *et al.* 2017). Metabolite profiling helped to decipher the role of chitosan, a natural linear polysaccharide, in increasing the growth of wheat seedlings through enhanced the metabolism of Carbon and Nitrogen assimilation (Zhang *et al.* 2017). Allwood *et al.*, (2015) studied the metabolite profiles and showed the effect of nitrate deprivation on the composition of amino acids, organic acids and carbohydrates in wheat. Sixty-four metabolites display differential expression in the heat-stressed pollen of wheat (Thomason *et al.* 2018). Enzymes involved in glycolysis, TCA cycle, and lipid metabolism reduced in the wheat grains exposed to heat stress (Wang *et al.* 2018). Combing both metabolomic and protein profiling of wheat grains showed the insights on high temperature stress adaption through channelling of photosynthates towards synthesis of heat responsive proteins in wheat (Wang *et al.* 2018).

## Ionomics

Ionomics harness the potential of the high-throughput technologies for profiling the elements in order to study the molecular mechanism underlying nutrient and trace element composition in plant tissues (Huang and Salt 2016). Fatiukha *et al.*, (2020) studied the association of ionome comprising 11 elements *viz.*, Aluminum, (Al), Calcium, (Ca), Copper, (Cu), Iron, (Fe), Potassium, (K), Magnesium, (Mg), Manganese, (Mn), Phosphorus, (P), Sulfur, (S) and Zinc, (Zn) linked to grain development in wheat. Interestingly, 617 QTL effects distributed among 105 QT loci were known to regulate the grain ionome in wheat. Three QTLs *viz.*, 3A.3, 5B.4 and 7B.1 display strong effects with Ca concentration codes for the genes TRIDC3AG040050, a calcium channel protein, TRIDC5BG045550 (Ca-transporting ATPase), and TRIDC7BG001820, $Ca^+$ exchanger protein, respectively, whereas 2B.6, 5A.1 and 7A.6 loci linked to Cu contains three Cu-transporting ATPase genes (TRIDC2BG062440, TRIDC5AG004320 and TRIDC7AG058420). The QTL 6A.3 region codes for three genes related to the K transporter family (TRIDC6AG029740, TRIDC6AG044480 and TRIDC6AG044820) (Fatiukha *et al.* 2020).

## Functional Genomics

Decoding of the genomes of important plants including the model plants rice and Arabidopsis have helped in identification and characterisation of many agronomically important genes through functional genomics approach (Bevan 2005; Li *et al.* 2018b; Lenka *et al.* 2018; Muthusamy *et al.* 2019). Thus, availability of chromosome-based high-quality draft genome sequences of wheat would be helpful in identification of genes and its regulatory regions linked important traits (Muthusamy *et al.* 2017; The International Wheat Genome Sequencing Consortium (IWGSC) *et al.* 2018). The tissue- and stress-specific transcriptome datasets can be used to understand the similarities and differences between homoeologous chromosomes (Muthusamy *et al.* 2016; Ramirez *et al.* 2018). Recent studies have shown, the major role of epigenetic inheritance, including DNA methylation, histone modification, RNA mediated silencing and chromatin modification in thermo-tolerance (Friedrich *et al.* 2019). Transgenic tobacco lines overexpressing a wheat F-box protein *TaFBA1* displayed enhanced high temperature stress tolerance (Li *et al.* 2018a). Similarly, the role of glucose in glucose-regulated *HLP1* homeostasis mechanism in regulating the thermo-memory in Arabidopsis (Sharma *et al.* 2019).

## Genetic Engineering

Heat shock transcription factors and heat shock proteins (HSPs) play major role in protecting plants under high temperature stress conditions, could be a potential target for engineering thermotolerance in wheat (Fragkostefanakis *et al.* 2015; Muthusamy *et al.* 2017). Availability of efficient transformation and regeneration protocol in wheat could be exploited for functional characterisation of the genes (Abouseadaa *et al.* 2015). Overexpression of *TaPEPKR2*, a *phosphoenolpyruvate carboxylase kinase-related kinase* gene enhances the high temperature and dehydration tolerance in transgenic wheat lines (Zang *et al.* 2018).

## Genomics Assisted Breeding

Integration of modern genomics tools in conventional breeding programs enhances the efficiency of selection by reduces the breeding cycle with minimal resources (Kole *et al.* 2015; Leng *et al.* 2017). Genomic selection (GS) is considered to be potential strategy for enhancing the complex traits in the polyploids like wheat (Kumar *et al.* 2018). The phenotyping datasets available in the large coordinated programs would be exploited for identification of QTLs associated with thermotolerance and building models for predicting the breeding values of progenies (Norman *et al.* 2018). Norman and co-workers (2018) studied a panel of 10,375 bread wheat using 18,101 SNP markers and showed the influence of several factors including the type of training population, population size, statistical models and marker density affecting the prediction accuracy. Integration of genomic selection in selection process resulted in 10% gain for grain yield and agronomic traits in soft red winter wheat (Lozada *et al.* 2019). Thus, factors affecting the prediction accuracy need to be considered for harnessing the full potential of genomic selection in wheat (Kumar *et al.* 2018; Norman *et al.* 2018).

## Integration of Omics Tools for Enhancing Thermotolerance

The large collections of mutant lines and advanced omics technologies can be exploited for deciphering the molecular mechanism imparting thermo-tolerance in wheat (Chen *et al.* 2012; Krasileva *et al.* 2017). Development of tissue-specific and developmental stage specific transcriptome, proteome, metabolome and ionome profiles under heat stress conditions coupled with functional analysis of genes would helpful in characterization of mode-of-action of the candidate genes(Lenka *et al.* 2019). This will aid in development of most relevant molecular markers for developing high temperature stress tolerant and high yielding cultivars. Thus, integration of several omics approaches including transcriptomic, proteomic and metabolomic datasets under high temperature stress conditions would facilitate understanding of the gene regulatory and cellular networks imparting thermo-tolerance in wheat (Thomason *et al.* 2018; Jagannadham *et al.* 2019). Thus, utilization of genome sequence information for development of molecular markers in wheat breeding programs for mapping QTLs/genes lined to agronomically important traits. Further, functional characterization of key genes through functional genomics and introgression of the superior genes/alleles through marker assisted-breeding would enhance the thermotolerance in wheat.

## CONCLUSION

High temperature stress due to climate change is becoming noticeable globally and a threat to wheat production. Understanding plant response mechanisms to high temperature stress and devising suitable mitigation strategies would help in minimizing yield losses due to stress. Practicing an integrated approach including developing tolerant cultivars, managing the crop by agronomic management options such as increased irrigation frequency, residue mulching and foliar spray of micronutrients involved in conserving water are the key to sustaining yields at farmer's field. At the research front recent developments in the field of omics like transcriptomics, proteomics, metabolomics and ionomics provide opportunities to manipulate molecular and physiological mechanisms or pathways governing component traits controlling high temperature tolerance. The aim should be in integrating the omics tools into conventional breeding strategies, in order to speed up the release of tolerant cultivars backed by experience of large scale testing at the hot spot locations.

## CONSENT FOR PUBLICATION

Not applicable.

## CONFLICT OF INTEREST

The authors confirm that this chapter content has no conflict of interest.

## ACKNOWLEDGEMENTS

Declared none.

## REFERENCES

Abouseadaa, H.H., Osman, G.H., Ramadan, A.M., Hassanein, S.E., Abdelsattar, M.T., Morsy, Y.B., Alameldin, H.F., El-Ghareeb, D.K., Nour-Eldin, H.A., Salem, R., Gad, A.A., Elkhodary, S.E., Shehata, M.M., Mahfouz, H.M., Eissa, H.F., Bahieldin, A. (2015). Development of transgenic wheat (*Triticum aestivum* L.) expressing avidin gene conferring resistance to stored product insects. *BMC Plant Biol., 15*, 183.
[http://dx.doi.org/10.1186/s12870-015-0570-x] [PMID: 26194497]

Al-Karaki, G.N. (2012). Phenological development-yield relationships in durum wheat cultivars under late-season high-temperature stress in a semiarid environment. *ISRN Agronomy, 2012*, 1-7.
[http://dx.doi.org/10.5402/2012/456856]

Al-Khatib, K., Paulsen, G.M. (1990). Photosynthesis and productivity during high-temperature stress of wheat genotypes from major world regions. *Crop Sci., 30*, 1127.
[http://dx.doi.org/10.2135/cropsci1990.0011183X003000050034x]

Allakhverdiev, S.I., Kreslavski, V.D., Klimov, V.V., Los, D.A., Carpentier, R., Mohanty, P. (2008). Heat stress: an overview of molecular responses in photosynthesis. *Photosynth. Res., 98*(1-3), 541-550.
[http://dx.doi.org/10.1007/s11120-008-9331-0] [PMID: 18649006]

Allwood, J.W., Chandra, S., Xu, Y., Dunn, W.B., Correa, E., Hopkins, L., Goodacre, R., Tobin, A.K., Bowsher, C.G. (2015). Profiling of spatial metabolite distributions in wheat leaves under normal and nitrate limiting conditions. *Phytochemistry, 115*, 99-111.
[http://dx.doi.org/10.1016/j.phytochem.2015.01.007] [PMID: 25680480]

Appels, R., Eversole, K., Stein, N., Feuillet, C., Keller, B., Rogers, J., Ronen, G. (2018). Shifting the limits in wheat research and breeding using a fully annotated reference genome. *Science, 361*(6403).
[http://dx.doi.org/10.1126/science.aar7191] [PMID: 30115783]

Aslam, B., Basit, M., Nisar, M.A., Khurshid, M., Rasool, M.H. (2017). Proteomics: technologies and their applications. *J. Chromatogr. Sci., 55*(2), 182-196.
[http://dx.doi.org/10.1093/chromsci/bmw167] [PMID: 28087761]

Asseng, S., van Herwaarden, A.F. (2003). Analysis of the benefits to wheat yield from assimilates stored prior to grain filling in a range of environments. *Plant Soil, 256*, 217-229.
[http://dx.doi.org/10.1023/A:1026231904221]

Asthir, B., Koundal, A., Bains, N.S. (2012). Putrescine modulates antioxidant defense response in wheat under high temperature stress. *Biol. Plant., 56*, 757-761.
[http://dx.doi.org/10.1007/s10535-012-0209-1]

Awasthi, R., Kaushal, N., Vadez, V., Turner, N.C., Berger, J., Siddique, K.H.M., Nayyar, H. (2014). Individual and combined effects of transient drought and heat stress on carbon assimilation and seed filling in

chickpea. *Funct. Plant Biol., 41*(11), 1148-1167.
[http://dx.doi.org/10.1071/FP13340] [PMID: 32481065]

Awlachew, Z.T., Singh, R., Kaur, S., Bains, N.S., Chhuneja, P. (2016). Transfer and mapping of the heat tolerance component traits of aegilops speltoides in tetraploid wheat triticum durum. *Mol. Breed., 36.*
[http://dx.doi.org/10.1007/s11032-016-0499-2]

Badaruddin, M., Reynolds, M.P., Ageeb, O.A.A. (1999). Wheat management in warm environments: effect of organic and inorganic fertilizers, irrigation frequency, and mulching. *Agron. J., 91*, 975-983.
[http://dx.doi.org/10.2134/agronj1999.916975x]

Bala, P., Sikder, S. (2017). Evaluation of heat tolerance of wheat genotypes through membrane thermostability test. *MAYFEB J. Agricult. Sci., 2,* 1-6.

Banowetz, G.M., Azevedo, M.D., Stout, R. (2008). Morphological adaptations of hot springs panic grass (Dichanthelium lanigunosum var sericeum (Schmoll) to thermal stress. *J. Therm. Biol., 33,* 106-116.
[http://dx.doi.org/10.1016/j.jtherbio.2007.08.006]

Bennett, D., Reynolds, M., Mullan, D., Izanloo, A., Kuchel, H., Langridge, P., Schnurbusch, T. (2012). Detection of two major grain yield QTL in bread wheat (*Triticum aestivum* L.) under heat, drought and high yield potential environments. *Theor. Appl. Genet., 125*(7), 1473-1485.
[http://dx.doi.org/10.1007/s00122-012-1927-2] [PMID: 22772727]

Bevan, M., Walsh, S. (2005). The Arabidopsis genome: a foundation for plant research. *Genome Res., 15*(12), 1632-1642.
[http://dx.doi.org/10.1101/gr.3723405] [PMID: 16339360]

Blum, A. (2018). *Plant Breeding for Stress Environments.* CRC press.
[http://dx.doi.org/10.1201/9781351075718]

Blum, A., Ebercon, A. (1981). Cell membrane stability as a measure of drought and heat tolerance in wheat 1. *Crop Sci., 21*, 43-47.
[http://dx.doi.org/10.2135/cropsci1981.0011183X002100010013x]

Blum, A., Klueva, N., Nguyen, H.T. (2001). Wheat cellular thermotolerance is related to yield under heat stress. *Euphytica, 117*, 117-123.
[http://dx.doi.org/10.1023/A:1004083305905]

Borrill, P., Ramirez-Gonzalez, R., Uauy, C. (2016). expVIP: a customizable RNA-seq data analysis and visualization platform. *Plant Physiol., 170*(4), 2172-2186.
[http://dx.doi.org/10.1104/pp.15.01667] [PMID: 26869702]

Britz, S.J., Prasad, P.V.V., Moreau, R.A., Allen, L.H., Jr, Kremer, D.F., Boote, K.J. (2007). Influence of growth temperature on the amounts of tocopherols, tocotrienols, and γ-oryzanol in brown rice. *J. Agric. Food Chem., 55*(18), 7559-7565.
[http://dx.doi.org/10.1021/jf0637729] [PMID: 17725318]

Cargnin, A., Souza, M., Dias, D., Machado, J.C., Machado, C.G., Sofiatti, V. (2006). Tolerância ao estresse de calor em genótipos de trigo na fase de germinação. *Bragantia, 65*, 245-251.
[http://dx.doi.org/10.1590/S0006-87052006000200006]

Chauhan, H., Khurana, N., Tyagi, A.K., Khurana, J.P., Khurana, P. (2011). Identification and characterization of high temperature stress responsive genes in bread wheat (*Triticum aestivum* L.) and their regulation at various stages of development. *Plant Mol. Biol., 75*(1-2), 35-51.
[http://dx.doi.org/10.1007/s11103-010-9702-8] [PMID: 20972607]

Chen, L., Huang, L., Min, D., Phillips, A., Wang, S., Madgwick, P.J., Parry, M.A.J., Hu, Y-G. (2012). Development and characterization of a new TILLING population of common bread wheat (*Triticum aestivum* L.). *PLoS One, 7*(7), , e41570..
[http://dx.doi.org/10.1371/journal.pone.0041570] [PMID: 22844501]

Chen, X., Hu, A., Li, M., Shen, H., Mao, D. (2007). Oxidation failure of lead frame copper alloys with surface electroplated pure Cu. *In 2007 8th International Conference on Electronic Packaging Technology.,*

IEEE.1-5.
[http://dx.doi.org/10.1109/ICEPT.2007.4441545]

Consortium, I.W.G.S. & others. (2018). Shifting the limits in wheat research and breeding using a fully annotated reference genome. *Science, 361*(6403), eaar7191.
[http://dx.doi.org/10.1126/science.aar7191]

Cooper, P., Rao, K., Singh, P., Dimes, J., Traore, P., Rao, K., Dixit, P., Twomlow, S. (2009). Farming with current and future climate risk: Advancing a 'Hypothesis of Hope' for rainfed agriculture in the semi-arid tropics. *J. SAT Agric. Res., 7*, 1-19.

Cossani, C.M., Reynolds, M.P. (2012). Physiological traits for improving heat tolerance in wheat. *Plant Physiol., 160*(4), 1710-1718.
[http://dx.doi.org/10.1104/pp.112.207753] [PMID: 23054564]

Dash, S., Mohanty, N. (2001). Evaluation of assays for the analysis of thermo-tolerance and recovery potentials of seedlings of wheat (*Triticum aestivum* L.) cultivars. *J. Plant Physiol., 158*, 1153-1165.
[http://dx.doi.org/10.1078/0176-1617-00243]

de Leonardis, A.M., Fragasso, M., Beleggia, R., Ficco, D.B., de Vita, P., Mastrangelo, A.M. (2015). Effects of heat stress on metabolite accumulation and composition, and nutritional properties of durum wheat grain. *Int. J. Mol. Sci., 16*(12), 30382-30404.
[http://dx.doi.org/10.3390/ijms161226241] [PMID: 26703576]

Dias, A., Lidon, F. (2010). Bread and durum wheat tolerance under heat stress: A synoptical overview. *Emir. J. Food Agric., 22*, 412.
[http://dx.doi.org/10.9755/ejfa.v22i6.4660]

Ding, H., Han, Q., Ma, D., Hou, J., Huang, X., Wang, C., Xie, Y., Kang, G., Guo, T. (2018). Characterizing physiological and proteomic analysis of the action of H2S to mitigate drought stress in young seedling of wheat. *Plant Mol. Biol. Report., 36*, 45-57.
[http://dx.doi.org/10.1007/s11105-017-1055-x]

Dupont, F.M., Altenbach, S.B. (2003). Molecular and biochemical impacts of environmental factors on wheat grain development and protein synthesis. *J. Cereal Sci., 38*, 133-146.
[http://dx.doi.org/10.1016/S0733-5210(03)00030-4]

Dwivedi, S., Kumar, G., Basu, S., Kumar, S., Rao, K., Choudhary, A. (1991). Physiological and molecular aspects of heat tolerance in wheat. *SABRAO Journal of Breeding & Genetics, 50*.

Dwyer, L., Tollenaar, M., Houwing, L. (2018). A nondestructive method to monitor leaf greenness in corn. *Can. J. Plant Sci., 71*, 505-509.

Ebeed, H.T. (2019). Omics approaches for developing abiotic stress tolerance in wheat. *Wheat Production in Changing Environments* (pp. 443-463). Singapore: Springer.
[http://dx.doi.org/10.1007/978-981-13-6883-7_17]

Ehdaie, B., Waines, J. (1992). Heat resistance in wild triticum and aegilops. *J. Genet. Breed., 46*, 221-227.

ElBasyoni, I., Saadalla, M., Baenziger, S., Bockelman, H., Morsy, S. (2017). Cell hembrane stability and association mapping for drought and heat tolerance in a worldwide wheat collection. *Sustainability, 9*, 1606.
[http://dx.doi.org/10.3390/su9091606]

Essemine, J, Ammar, S, Bouzid, S (2010). Impact of heat stress on germination and growth in higher plants: Physiological, biochemical and molecular repercussions and mechanisms of defence. *J. Biol. Sci., 10*, 565-572.
[http://dx.doi.org/10.3923/jbs.2010.565.572]

Farooq, M., Gogoi, N., Barthakur, S., Baroowa, B., Bharadwaj, N., Alghamdi, S.S., Siddique, K.H.M. (2017). Drought stress in G]grain legumes during reproduction and grain filling. *J. Agron. Crop Sci., 203*, 81-102.
[http://dx.doi.org/10.1111/jac.12169]

Fatiukha, A., Klymiuk, V., Peleg, Z., Saranga, Y., Cakmak, I., Krugman, T., Korol, A.B., Fahima, T. (2020).

Variation in phosphorus and sulfur content shapes the genetic architecture and phenotypic associations within the wheat grain ionome. *Plant J., 101*(3), 555-572.
[http://dx.doi.org/10.1111/tpj.14554] [PMID: 31571297]

Fokar, M., Nguyen, H., Blum, A. (1998). Heat tolerance in spring wheat. I. Genetic variability and heritability of cellular thermotolerance. *Euphytica, 104*, 1-8.
[http://dx.doi.org/10.1023/A:1018346901363]

Fragkostefanakis, S., Röth, S., Schleiff, E., Scharf, K-D. (2015). Prospects of engineering thermotolerance in crops through modulation of heat stress transcription factor and heat shock protein networks. *Plant Cell Environ., 38*(9), 1881-1895.
[http://dx.doi.org/10.1111/pce.12396] [PMID: 24995670]

Friedrich, T., Faivre, L., Bäurle, I., Schubert, D. (2019). Chromatin-based mechanisms of temperature memory in plants. *Plant Cell Environ., 42*(3), 762-770.
[http://dx.doi.org/10.1111/pce.13373] [PMID: 29920687]

Gebbing, T., Schnyder, H. (1999). Pre-anthesis reserve utilization for protein and carbohydrate synthesis in grains of wheat. *Plant Physiol., 121*(3), 871-878.
[http://dx.doi.org/10.1104/pp.121.3.871] [PMID: 10557235]

Goff, S.A., Ricke, D., Lan, T-H., Presting, G., Wang, R., Dunn, M., Glazebrook, J., Sessions, A., Oeller, P., Varma, H., Hadley, D., Hutchison, D., Martin, C., Katagiri, F., Lange, B.M., Moughamer, T., Xia, Y., Budworth, P., Zhong, J., Miguel, T., Paszkowski, U., Zhang, S., Colbert, M., Sun, W.L., Chen, L., Cooper, B., Park, S., Wood, T.C., Mao, L., Quail, P., Wing, R., Dean, R., Yu, Y., Zharkikh, A., Shen, R., Sahasrabudhe, S., Thomas, A., Cannings, R., Gutin, A., Pruss, D., Reid, J., Tavtigian, S., Mitchell, J., Eldredge, G., Scholl, T., Miller, R.M., Bhatnagar, S., Adey, N., Rubano, T., Tusneem, N., Robinson, R., Feldhaus, J., Macalma, T., Oliphant, A., Briggs, S. (2002). A draft sequence of the rice genome (*Oryza sativa* L. ssp. *japonica*). *Science, 296*(5565), 92-100.
[http://dx.doi.org/10.1126/science.1068275] [PMID: 11935018]

Großkinsky, D.K., Syaifullah, S.J., Roitsch, T. (2018). Integration of multi-omics techniques and physiological phenotyping within a holistic phenomics approach to study senescence in model and crop plants. *J. Exp. Bot., 69*(4), 825-844.
[http://dx.doi.org/10.1093/jxb/erx333] [PMID: 29444308]

Guan, Y., Li, G., Chu, Z., Ru, Z., Jiang, X., Wen, Z., Zhang, G., Wang, Y., Zhang, Y., Wei, W. (2019). Transcriptome analysis reveals important candidate genes involved in grain-size formation at the stage of grain enlargement in common wheat cultivar "Bainong 4199". *PLoS One, 14*(3), , e0214149..
[http://dx.doi.org/10.1371/journal.pone.0214149] [PMID: 30908531]

Gupta, N.K., Agarwal, S., Agarwal, V.P., Nathawat, N.S., Gupta, S., Singh, G. (2013). Effect of short-term heat stress on growth, physiology and antioxidative defence system in wheat seedlings. *Acta Physiol. Plant., 35*, 1837-1842.
[http://dx.doi.org/10.1007/s11738-013-1221-1]

Hall, A.E. (1993). Physiology and breeding for heat tolerance in cowpea, and comparison with other crops. *Adapt. Food Crop Temp. Water Stress, 271-284.*

Hall, A.E. (1992). Breeding for heat tolerance. *Plant Breed. Rev., 10*, 129-168.

Hasanuzzaman, M., Mahmud, J.A., Anee, T.I., Nahar, K., Islam, M.T. (2018). Drought Stress Tolerance in Wheat: Omics Approaches in Understanding and Enhancing Antioxidant Defense. *Abiotic Stress-Mediated Sensing and Signaling in Plants: An Omics Perspective.* (pp. 267-307). Springer Singapore, Singapore.
[http://dx.doi.org/10.1007/978-981-10-7479-0_10]

Huang, X-Y., Salt, D.E. (2016). Plant ionomics: from elemental profiling to environmental adaptation. *Mol. Plant, 9*(6), 787-797.
[http://dx.doi.org/10.1016/j.molp.2016.05.003] [PMID: 27212388]

Hurkman, W.J., McCue, K.F., Altenbach, S.B., Korn, A., Tanaka, C.K., Kothari, K.M., Johnson, E.L., Bechtel, D.B., Wilson, J.D., Anderson, O.D., DuPont, F.M. (2003). Effect of temperature on expression of genes encoding enzymes for starch biosynthesis in developing wheat endosperm. *Plant Sci., 164*, 873-881. [http://dx.doi.org/10.1016/S0168-9452(03)00076-1]

Ibrahim, A.M., Quick, J.S. (2001). Heritability of heat tolerance in winter and spring wheat. *Crop Sci., 41*, 1401-1405. [http://dx.doi.org/10.2135/cropsci2001.4151401x]

ICAR-IIWBR. (2018). Progress Report of AICRP on Wheat and Barley 2017-18, Crop Improvement. Ravish, Chatrath, Vinod, Tiwari, Gyanendra, Singh, Ratan, Tiwari, BS, Tyagi, Arun, Gupta, Raj, Kumar, SK, Singh, Lokendra, Kumar, AK, Sharma, Hanif, Khan, Satish, Kumar, Charan, Singh, CN, Mishra, K, Venkatesh, Mamrutha, HM, Vikas, Gupta, Rinki, , Gopalareddy, K, Ajay, Verma, GP, Singh (p. 206). Karnal, Haryana, India: ICAR - Indian Institute of Wheat and Barley Research. http://www.iiwbr.org

IPCC. (2007). *Climate Change 2007: The Physical Science Basis: Contribution of Working Group I to the 4ᵗʰ Assessment Report of the Intergovernmental Panel on Climate Change.* Cambridge University Press.

Jacob, P., Hirt, H., Bendahmane, A. (2017). The heat-shock protein/chaperone network and multiple stress resistance. *Plant Biotechnol. J., 15*(4), 405-414. [http://dx.doi.org/10.1111/pbi.12659] [PMID: 27860233]

Jagannadham, P.T.K., Muthusamy, S.K., Chidambaranathan, P. (2019). Micromics: A Novel Approach to Understand the Molecular Mechanisms in Plant Stress Tolerance. *Recent Approaches in Omics for Plant Resilience to Climate Change.* (pp. 93-108). Cham: Springer International Publishing. [http://dx.doi.org/10.1007/978-3-030-21687-0_5]

Jain, N., Singh, G.P., Singh, P.K., Ramya, P., Krishna, H., Ramya, K.T., Todkar, L., Amasiddha, B., Kumar, K.C.P., Vijay, P., Jadon, V., Dutta, S., Rai, N., Sinha, N., Prabhu, K.V. (2014). Molecular approaches for wheat improvement under drought and heat stress. *Indian J. Genet. Plant Breed., 74*, 578. [http://dx.doi.org/10.5958/0975-6906.2014.00893.1]

Joshi, A., Mishra, B., Chatrath, R., Ferrara, G.O., Singh, R.P. (2007). Wheat improvement in India: present status, emerging challenges and future prospects. *Euphytica, 157*, 431-446. a [http://dx.doi.org/10.1007/s10681-007-9385-7]

Joshi, A.K., Ortiz-Ferrara, G., Crossa, J., Singh, G., Sharma, R.C., Chand, R., Parsad, R. (2007). Combining superior agronomic performance and terminal heat tolerance with resistance to spot blotch (Bipolaris sorokiniana) of wheat in the warm humid Gangetic Plains of South Asia. *Field Crops Res., 103*, 53-61. b [http://dx.doi.org/10.1016/j.fcr.2007.04.010]

Katiyar, A., Smita, S., Muthusamy, S.K., Chinnusamy, V., Pandey, D.M., Bansal, K.C. (2015). Identification of novel drought-responsive microRNAs and trans-acting siRNAs from Sorghum bicolor (L.) Moench by high-throughput sequencing analysis. *Front. Plant Sci., 6*, 506. [http://dx.doi.org/10.3389/fpls.2015.00506] [PMID: 26236318]

Kaul, S., Koo, H.L., Jenkins, J., Rizzo, M., Rooney, T., Tallon, L.J. (2000). Analysis of the genome sequence of the flowering plant *Arabidopsis thaliana. Nature, 408*(6814), 796-815.

Kaushal, N., Bhandari, K., Siddique, K.H., Nayyar, H. (2016). Food crops face rising temperatures: an overview of responses, adaptive mechanisms, and approaches to improve heat tolerance. *Cogent Food Agric., 2*, 1134380. [http://dx.doi.org/10.1080/23311932.2015.1134380]

Kole, C., Muthamilarasan, M., Henry, R., Edwards, D., Sharma, R., Abberton, M., Batley, J., Bentley, A., Blakeney, M., Bryant, J., Cai, H., Cakir, M., Cseke, L.J., Cockram, J., de Oliveira, A.C., De Pace, C., Dempewolf, H., Ellison, S., Gepts, P., Greenland, A., Hall, A., Hori, K., Hughes, S., Humphreys, M.W., Iorizzo, M., Ismail, A.M., Marshall, A., Mayes, S., Nguyen, H.T., Ogbonnaya, F.C., Ortiz, R., Paterson, A.H., Simon, P.W., Tohme, J., Tuberosa, R., Valliyodan, B., Varshney, R.K., Wullschleger, S.D., Yano, M., Prasad, M. (2015). Application of genomics-assisted breeding for generation of climate resilient crops: progress and prospects. *Front. Plant Sci., 6*, 563. [http://dx.doi.org/10.3389/fpls.2015.00563] [PMID: 26322050]

Krasileva, K.V., Vasquez-Gross, H.A., Howell, T., Bailey, P., Paraiso, F., Clissold, L., Simmonds, J.,

Ramirez-Gonzalez, R.H., Wang, X., Borrill, P., Fosker, C., Ayling, S., Phillips, A.L., Uauy, C., Dubcovsky, J. (2017). Uncovering hidden variation in polyploid wheat. *Proc. Natl. Acad. Sci. USA, 114*(6), E913-E921.
[http://dx.doi.org/10.1073/pnas.1619268114] [PMID: 28096351]

Kumar, R., Bohra, A., Pandey, A.K., Pandey, M.K., Kumar, A. (2017). Metabolomics for plant improvement: status and prospects. *Front. Plant Sci., 8*, 1302.
[http://dx.doi.org/10.3389/fpls.2017.01302] [PMID: 28824660]

Kumar, R.R., Sharma, S.K., Goswami, S., Singh, G.P., Singh, R., Singh, K., Pathak, H., Rai, R.D. (2013). *characterization of differentially expressed stress-associated proteins in starch granule development under heat stress in wheat (Triticum aestivum L.).* IJBB, *50*(2) [April 2013].

Kumar, S., Kaur, R., Kaur, N., Bhandhari, K., Kaushal, N., Gupta, K., Bains, T., Nayyar, H. (2011). Heat-stress induced inhibition in growth and chlorosis in mungbean (*Phaseolus aureus* Roxb.) is partly mitigated by ascorbic acid application and is related to reduction in oxidative stress. *Acta Physiol. Plant., 33*, 2091.
[http://dx.doi.org/10.1007/s11738-011-0748-2]

Kumar, S., Muthusamy, S.K., Mishra, C.N., Gupta, V., Venkatesh, K. (2018). Importance of genomic selection in crop improvement and future prospects. *Advanced Molecular Plant Breeding: Meeting the Challenge of Food Security.* CRC Press.
[http://dx.doi.org/10.1201/b22473-8]

Lamaoui, M., Jemo, M., Datla, R., Bekkaoui, F. (2018). Heat and drought stresses in crops and approaches for their mitigation. *Front Chem., 6*, 26.
[http://dx.doi.org/10.3389/fchem.2018.00026] [PMID: 29520357]

Lehman, V., Engelke, M.C. (1993). Heritability of creeping bentgrass shoot water content under soil dehydration and elevated temperatures. *Crop Sci., 33*, 1061-1066.
[http://dx.doi.org/10.2135/cropsci1993.0011183X003300050037x]

Leng, P., Lübberstedt, T., Xu, M. (2017). Genomics-assisted breeding – A revolutionary strategy for crop improvement. *J. Integr. Agric., 16*, 2674-2685.
[http://dx.doi.org/10.1016/S2095-3119(17)61813-6]

Lenka, S.K., Muthusamy, S.K., Chinnusamy, V., Bansal, K.C. (2018). Ectopic expression of rice PLY3 enhances cold and drought tolerance in *Arabidopsis thaliana. Mol. Biotechnol., 60*(5), 350-361.
[http://dx.doi.org/10.1007/s12033-018-0076-5] [PMID: 29574592]

Lenka, S.K., Singh, A.K., Muthusamy, S.K., Smita, S., Chinnusamy, V., Bansal, K.C. (2019). Heterologous expression of rice RNA-binding glycine-rich (RBG) gene OsRBGD3 in transgenic *Arabidopsis thaliana* confers cold stress tolerance. *Funct. Plant Biol., 46*(5), 482-491.
[http://dx.doi.org/10.1071/FP18241] [PMID: 30940336]

Li, Q., Wang, W., Wang, W., Zhang, G., Liu, Y., Wang, Y., Wang, W. (2018). Wheat F-Box protein gene *TaFBA1* is involved in plant tolerance to heat stress. *Front. Plant Sci., 9*, 521. a
[http://dx.doi.org/10.3389/fpls.2018.00521] [PMID: 29740462]

Li, Y., Xiao, J., Chen, L., Huang, X., Cheng, Z., Han, B., Zhang, Q., Wu, C. (2018). Rice functional genomics research: past decade and future. *Mol. Plant, 11*(3), 359-380. b
[http://dx.doi.org/10.1016/j.molp.2018.01.007] [PMID: 29409893]

Lim, P.O., Kim, H.J., Nam, H.G. (2007). Leaf senescence. *Annu. Rev. Plant Biol., 58*, 115-136.
[http://dx.doi.org/10.1146/annurev.arplant.57.032905.105316] [PMID: 17177638]

Liu, B., Liu, L., Asseng, S., Zou, X., Li, J., Cao, W., Zhu, Y. (2016). Modelling the effects of heat stress on post-heading durations in wheat: A comparison of temperature response routines. *Agric. For. Meteorol., 222*, 45-58.
[http://dx.doi.org/10.1016/j.agrformet.2016.03.006]

Lobell, D.B., Field, C.B. (2007). Global scale climate–crop yield relationships and the impacts of recent warming. *Environ. Res. Lett., 2*, , 014002..
[http://dx.doi.org/10.1088/1748-9326/2/1/014002]

Lobell, D.B., Sibley, A., Ivan Ortiz-Monasterio, J. (2012). Extreme heat effects on wheat senescence in India. *Nat. Clim. Chang., 2*, 186-189.
[http://dx.doi.org/10.1038/nclimate1356]

Lopes, M.S., Reynolds, M.P. (2012). Stay-green in spring wheat can be determined by spectral reflectance measurements (normalized difference vegetation index) independently from phenology. *J. Exp. Bot., 63*(10), 3789-3798.
[http://dx.doi.org/10.1093/jxb/ers071] [PMID: 22412185]

Lozada, D.N., Mason, R.E., Sarinelli, J.M., Brown-Guedira, G. (2019). Accuracy of genomic selection for grain yield and agronomic traits in soft red winter wheat. *BMC Genet., 20*(1), 82.
[http://dx.doi.org/10.1186/s12863-019-0785-1] [PMID: 31675927]

Maes, B., Trethowan, R.M., Reynolds, M.P., van Ginkel, M., Skovmand, B. (2001). The influence of glume pubescence on spikelet temperature of wheat under freezing conditions. *Funct. Plant Biol., 28*, 141.
[http://dx.doi.org/10.1071/PP00049]

Mamrutha, H.M., Sharma, D., Kumar, K.S., Venkatesh, K., Tiwari, V., Sharma, I. (2017). Influence of diurnal irradiance variation on chlorophyll values in wheat: a comparative study using different chlorophyll meters. *Natl. Acad. Sci. Lett., 40*, 221-224.
[http://dx.doi.org/10.1007/s40009-017-0544-7]

Mason, R.E., Mondal, S., Beecher, F.W., Pacheco, A., Jampala, B., Ibrahim, A.M., Hays, D.B. (2010). QTL associated with heat susceptibility index in wheat (*Triticum aestivum* L.) under short-term reproductive stage heat stress. *Euphytica, 174*, 423-436.
[http://dx.doi.org/10.1007/s10681-010-0151-x]

Maxwell, K., Johnson, G.N. (2000). Chlorophyll fluorescence - a practical guide. *J. Exp. Bot., 51*(345), 659-668.
[http://dx.doi.org/10.1093/jexbot/51.345.659] [PMID: 10938857]

Meena, K.K., Sorty, A.M., Bitla, U.M., Choudhary, K., Gupta, P., Pareek, A., Singh, D.P., Prabha, R., Sahu, P.K., Gupta, V.K., Singh, H.B., Krishanani, K.K., Minhas, P.S. (2017). Abiotic stress responses and microbe-mediated mitigation in plants: the omics strategies. *Front. Plant Sci., 8*, 172.
[http://dx.doi.org/10.3389/fpls.2017.00172] [PMID: 28232845]

Meena, R.P., Karnam, V., Sendhil, R., Sharma, R., Tripathi, S., Singh, G.P. (2019). Identification of water use efficient wheat genotypes with high yield for regions of depleting water resources in India. *Agric. Water Manage., 223*, 105709.
[http://dx.doi.org/10.1016/j.agwat.2019.105709]

Meena, R.P., Karnam, V., Tripathi, S.C., Jha, A., Sharma, R.K., Singh, G.P. (2019). Irrigation management strategies in wheat for efficient water use in the regions of depleting water resources. *Agric. Water Manage., 214*, 38-46. b
[http://dx.doi.org/10.1016/j.agwat.2019.01.001]

Meena, R.P., Sharma, R., Chhokar, R., Chander, S., Tripathi, S., Kumar, R., Sharma, I. (2015). Improving water use efficiency of rice-wheat cropping system by adopting micro-irrigation systems. *Int. J. Bioresource Stress Manag., 6*.

Moffatt, J., Sears, R., Cox, T., Paulsen, G. (1990). Wheat high temperature tolerance during reproductive growth. II. Genetic analysis of chlorophyll fluorescence. *Crop Sci., 30*, 886-889.

Mondal, Suchismita. (2011). Defining the Molecular and Physiological Role of Leaf Cuticular Waxes in Reproductive Stage Heat Tolerane in Wheat. *Doctoral dissertation.* Texas A&M University. Available electronically from http : / /hdl .handle .net /1969 .1 /ETD -TAMU -2011 -05 -9449.

Mondal, S., Rutkoski, J.E., Velu, G., Singh, P.K., Crespo-Herrera, L.A., Guzmán, C., Bhavani, S., Lan, C., He, X., Singh, R.P. (2016). Harnessing diversity in wheat to enhance grain yield, climate resilience, disease and insect pest resistance and nutrition through conventional and modern breeding approaches. *Front. Plant Sci., 7*, 991.

[http://dx.doi.org/10.3389/fpls.2016.00991] [PMID: 27458472]

Mondal, S., Singh, R.P., Crossa, J., Huerta-Espino, J., Sharma, I., Chatrath, R., Singh, G.P., Sohu, V.S., Mavi, G.S., Sukuru, V.S.P., Kalappanavar, I.K., Mishra, V.K., Hussain, M., Gautam, N.R., Uddin, J., Barma, N.C.D., Hakim, A., Joshi, A.K. (2013). Earliness in wheat: A key to adaptation under terminal and continual high temperature stress in South Asia. *Field Crops Res., 151*, 19-26.
[http://dx.doi.org/10.1016/j.fcr.2013.06.015]

Mullan, D.J., Reynolds, M.P. (2010). Quantifying genetic effects of ground cover on soil water evaporation using digital imaging. *Funct. Plant Biol., 37*, 703-712.
[http://dx.doi.org/10.1071/FP09277]

Muthusamy, S.K., Dalal, M., Chinnusamy, V., Bansal, K.C. (2017). Genome-wide identification and analysis of biotic and abiotic stress regulation of small heat shock protein (HSP20) family genes in bread wheat. *J. Plant Physiol., 211*, 100-113.
[http://dx.doi.org/10.1016/j.jplph.2017.01.004] [PMID: 28178571]

Muthusamy, S.K., Dalal, M., Chinnusamy, V., Bansal, K.C. (2016). Differential regulation of genes coding for organelle and cytosolic clpatases under biotic and abiotic stresses in wheat. *Front. Plant Sci., 7*, 929.
[http://dx.doi.org/10.3389/fpls.2016.00929] [PMID: 27446158]

Muthusamy, S.K., Lenka, S.K., Katiyar, A., Chinnusamy, V., Singh, A.K., Bansal, K.C. (2019). Genome-wide identification and analysis of biotic and abiotic stress regulation of $c_4$ photosynthetic pathway genes in rice. *Appl. Biochem. Biotechnol., 187*(1), 221-238.
[http://dx.doi.org/10.1007/s12010-018-2809-0] [PMID: 29915917]

Nagarajan, S., Rane, J. (2000). Relationship of seedling traits with drought tolerance in spring wheat cultivars. *Indian J. Plant. Physiol., 5*, 264-270.

Nandha, A., Mehta, D., Tulsani, N., Umretiya, N., Delvadiya, N., Kachhadiya, H. (2019). Transcriptome analysis of response to heat stress in heat tolerance and heat susceptible wheat (*Triticum aestivum* L.) genotypes. *Journal of Pharmacognosy and Phytochemistry, 8*, 275-284.

Nataraja, K., Jacob, J. (1999). Clonal differences in photosynthesis in Hevea brasiliensis Müll. Arg. *Photosynthetica, 36*, 89.
[http://dx.doi.org/10.1023/A:1007070820925]

Ni, Z., Li, H., Zhao, Y., Peng, H., Hu, Z., Xin, M., Sun, Q. (2018). Genetic improvement of heat tolerance in wheat: Recent progress in understanding the underlying molecular mechanisms. *Crop J., 6*, 32-41.
[http://dx.doi.org/10.1016/j.cj.2017.09.005]

Norman, A., Taylor, J., Edwards, J., Kuchel, H. (2018). Optimising genomic selection in wheat: Effect of marker density, population size and population structure on prediction accuracy. *G3: Genes, Genomes, Genetics, 8*(9), 2889-2899.
[http://dx.doi.org/https://doi.org/10.1534/g3.118.200311]

Parry, M.A., Reynolds, M., Salvucci, M.E., Raines, C., Andralojc, P.J., Zhu, X-G., Price, G.D., Condon, A.G., Furbank, R.T. (2011). Raising yield potential of wheat. II. Increasing photosynthetic capacity and efficiency. *J. Exp. Bot., 62*(2), 453-467.
[http://dx.doi.org/10.1093/jxb/erq304] [PMID: 21030385]

Pearce, S., Vazquez-Gross, H., Herin, S.Y., Hane, D., Wang, Y., Gu, Y.Q., Dubcovsky, J. (2015). WheatExp: an RNA-seq expression database for polyploid wheat. *BMC Plant Biol., 15*, 299.
[http://dx.doi.org/10.1186/s12870-015-0692-1] [PMID: 26705106]

Peng, D., Peng, H., Ni, Z., Nie, X., Yao, Y., Qin, D., He, K., Sun, Q. (2006). Heat stress-responsive transcriptome analysis of wheat by using GeneChip Barley1 Genome Array. *Prog. Nat. Sci., 16*, 1379-1387.

Pinto, R.S., Reynolds, M.P., Mathews, K.L., McIntyre, C.L., Olivares-Villegas, J-J., Chapman, S.C. (2010). Heat and drought adaptive QTL in a wheat population designed to minimize confounding agronomic effects. *Theor. Appl. Genet., 121*(6), 1001-1021.
[http://dx.doi.org/10.1007/s00122-010-1351-4] [PMID: 20523964]

Piramila, B., Prabha, A., Nandagopalan, V., Stanley, A. (2012). Effect of heat treatment on germination, seedling growth and some biochemical parameters of dry seeds of black gram. *Int. J. Pharm. Phytopharmacol. Res., 1*, 194-202.

Pittock, B., Arthington, A., Booth, T., Cowell, P., Hennessy, K., Howden, M., McMichael, T. (2003). Climate change: An Australian guide to the science and potential impacts. *Australian Greenhouse Office.*

Pradhan, G.P., Prasad, P.V.V., Fritz, A.K., Kirkham, M.B., Gill, B.S. (2012). High temperature tolerance in *Aegilops* species and its potential transfer to wheat. *Crop Sci., 52*, 292-304.
[http://dx.doi.org/10.2135/cropsci2011.04.0186]

Prasad, P.V.V., Djanaguiraman, M. (2014). Response of floret fertility and individual grain weight of wheat to high temperature stress: sensitive stages and thresholds for temperature and duration. *Funct. Plant Biol., 41*(12), 1261-1269.
[http://dx.doi.org/10.1071/FP14061] [PMID: 32481075]

Prasad, P.V.V., Pisipati, S.R., Momčilović, I., Ristic, Z. (2011). Independent and combined effects of high temperature and drought stress during grain filling on plant yield and chloroplast ef-tu expression in spring wheat: effects of temperature and drought on wheat plants. *J. Agron. Crop Sci., 197*, 430-441.
[http://dx.doi.org/10.1111/j.1439-037X.2011.00477.x]

Priya, M., Siddique, K.H.M., Dhankhar, O.P., Prasad, P.V.V., Hanumantha Rao, B., Nair, R.M., Nayyar, H. (2018). Molecular breeding approaches involving physiological and reproductive traits for heat tolerance in food crops. *Indian J. Plant. Physiol., 23*, 697-720.
[http://dx.doi.org/10.1007/s40502-018-0427-z]

Qaseem, M.F., Qureshi, R., Shaheen, H. (2019). Effects of pre-anthesis drought, heat and their combination on the growth, yield and physiology of diverse wheat (*Triticum aestivum* L.) Genotypes varying in sensitivity to heat and drought stress. *Sci. Rep., 9*(1), 6955.
[http://dx.doi.org/10.1038/s41598-019-43477-z] [PMID: 31061444]

Qin, D., Wu, H., Peng, H., Yao, Y., Ni, Z., Li, Z., Zhou, C., Sun, Q. (2008). Heat stress-responsive transcriptome analysis in heat susceptible and tolerant wheat (*Triticum aestivum* L.) by using Wheat Genome Array. *BMC Genomics, 9*, 432.
[http://dx.doi.org/10.1186/1471-2164-9-432] [PMID: 18808683]

Ramadas, S., Kiran Kumar, T.M., Pratap Singh, G. (2019). *Wheat Production in India: Trends and Prospects. In: Recent Advances in Grain Crops Research [Working Title]. IntechOpen.*

Ramirez, I.A., Abbate, P.E., Redi, I.W., Pontaroli, A.C. (2018). Effects of photoperiod sensitivity genes Ppd-B1 and Ppd-D1 on spike fertility and related traits in bread wheat. *Plant Breed., 137*, 320-325.
[http://dx.doi.org/10.1111/pbr.12585]

Ramírez-González, R.H., Borrill, P., Lang, D., Harrington, S.A., Brinton, J., Venturini, L., Davey, M., Jacobs, J., van Ex, F., Pasha, A., Khedikar, Y., Robinson, S.J., Cory, A.T., Florio, T., Concia, L., Juery, C., Schoonbeek, H., Steuernagel, B., Xiang, D., Ridout, C.J., Chalhoub, B., Mayer, K.F.X., Benhamed, M., Latrasse, D., Bendahmane, A., Wulff, B.B.H., Appels, R., Tiwari, V., Datla, R., Choulet, F., Pozniak, C.J., Provart, N.J., Sharpe, A.G., Paux, E., Spannagl, M., Bräutigam, A., Uauy, C. International Wheat Genome Sequencing Consortium. (2018). The transcriptional landscape of polyploid wheat. *Science, 361*(6403), , eaar6089..
[http://dx.doi.org/10.1126/science.aar6089] [PMID: 30115782]

Rangan, P., Furtado, A., Henry, R. (2020). Transcriptome profiling of wheat genotypes under heat stress during grain-filling. *J. Cereal Sci., 91*, , 102895..
[http://dx.doi.org/10.1016/j.jcs.2019.102895]

Ravichandran, S., Ragupathy, R., Edwards, T., Domaratzki, M., Cloutier, S. (2019). MicroRNA-guided regulation of heat stress response in wheat. *BMC Genomics, 20*(1), 488.
[http://dx.doi.org/10.1186/s12864-019-5799-6] [PMID: 31195958]

Reynolds, M. (2001). *Application of physiology in wheat breeding.* Cimmyt.

Reynolds, M., Foulkes, M.J., Slafer, G.A., Berry, P., Parry, M.A.J., Snape, J.W., Angus, W.J. (2009). Raising yield potential in wheat. *J. Exp. Bot., 60*(7), 1899-1918.
[http://dx.doi.org/10.1093/jxb/erp016] [PMID: 19363203]

Reynolds, M.P., Pierre, C.S., Saad, A.S., Vargas, M., Condon, A.G. (2007). Evaluating potential genetic gains in wheat associated with stress-adaptive trait expression in elite genetic resources under drought and heat stress. *Crop Sci., 47*, S–172..
[http://dx.doi.org/10.2135/cropsci2007.10.0022IPBS]

Richards, R., Lukacs, Z. (2002). Seedling vigour in wheat-sources of variation for genetic and agronomic improvement. *Aust. J. Agric. Res., 53*, 41-50.
[http://dx.doi.org/10.1071/AR00147]

Ristic, Z., Bukovnik, U., Prasad, P.V. (2007). Correlation between heat stability of thylakoid membranes and loss of chlorophyll in winter wheat under heat stress. *Crop Sci., 47*, 2067-2073.
[http://dx.doi.org/10.2135/cropsci2006.10.0674]

Rodríguez, M., Canales, E., Borrás-Hidalgo, O. (2005). Molecular aspects of abiotic stress in plants. *Biotecnol. Apl., 22*, 1-10.

Saadalla, M., Shanahan, J., Quick, J. (1990). Heat tolerance in winter wheat: I. Hardening and genetic effects on membrane thermostability. *Crop Sci., 30*, 1243-1247.
[http://dx.doi.org/10.2135/cropsci1990.0011183X003000060017x]

Salman, M., Majeed, S., Rana, I.A., Atif, R.M., Azhar, M.T. (2019). Novel Breeding and Biotechnological Approaches to Mitigate the Effects of Heat Stress on Cotton. *Recent Approaches in Omics for Plant Resilience to Climate Change.* Springer.
[http://dx.doi.org/10.1007/978-3-030-21687-0_11]

Sareen, S., Tyagi, B.S., Sharma, I. (2012). Response estimation of wheat synthetic lines to terminal heat stress using stress indices. *J. Agric. Sci., 4*.
[http://dx.doi.org/10.5539/jas.v4n10p97]

Sarieva, G., Kenzhebaeva, S., Lichtenthaler, H. (2010). Adaptation potential of photosynthesis in wheat cultivars with a capability of leaf rolling under high temperature conditions. *Russ. J. Plant Physiol., 57*, 28-36.
[http://dx.doi.org/10.1134/S1021443710010048]

Schmutz, J., Cannon, S.B., Schlueter, J., Ma, J., Mitros, T., Nelson, W., Xu, D. (2010). Genome sequence of the palaeopolyploid soybean. *Nature, 463*(7278), 178-183.

Schnable, P.S., Ware, D., Fulton, R.S., Stein, J.C., Wei, F., Pasternak, S., Liang, C., Zhang, J., Fulton, L., Graves, T.A. (2009). The B73 maize genome: complexity, diversity, and dynamics. *Science, 326*, 1112-1115.

Sehgal, A., Sita, K., Siddique, K.H.M., Kumar, R., Bhogireddy, S., Varshney, R.K., HanumanthaRao, B., Nair, R.M., Prasad, P.V.V., Nayyar, H. (2018). Drought or/and heat-stress effects on seed filling in food crops: impacts on functional biochemistry, seed yields, and nutritional quality. *Front. Plant Sci., 9*, 1705.
[http://dx.doi.org/10.3389/fpls.2018.01705] [PMID: 30542357]

Sevanto, S. (2014). Phloem transport and drought. *J. Exp. Bot., 65*(7), 1751-1759.
[http://dx.doi.org/10.1093/jxb/ert467] [PMID: 24431155]

Sharma, I., Tyagi, B., Singh, G., Venkatesh, K., Gupta, O. (2015). Enhancing wheat production-A global perspective. *Indian J. Agric. Sci., 85*, 3-13.

Sharma, M., Banday, Z.Z., Shukla, B.N., Laxmi, A. (2019). Glucose-regulated *HLP1* acts as a key molecule in governing thermomemory. *Plant Physiol., 180*(2), 1081-1100.
[http://dx.doi.org/10.1104/pp.18.01371] [PMID: 30890662]

Shepherd, T., Wynne Griffiths, D. (2006). The effects of stress on plant cuticular waxes. *New Phytol., 171*(3), 469-499.
[http://dx.doi.org/10.1111/j.1469-8137.2006.01826.x] [PMID: 16866954]

Srivastava, S., Pathak, A.D., Gupta, P.S., Shrivastava, A.K., Srivastava, A.K. (2012). Hydrogen peroxide-scavenging enzymes impart tolerance to high temperature induced oxidative stress in sugarcane. *J. Environ. Biol., 33*(3), 657-661.
[PMID: 23029918]

Sun, Q., Quick, J. (1991). Chromosomal locations of genes for heat tolerance in tetraploid wheat. *Cereal Res. Commun.,* 431-437.

Team, C.W., Pachauri, R., Meyer, L. (2014). IPCC, 2014: Climate change: Synthesis report. *Contribution of working groups I II III to fifth assess rep intergov panel clim chang IPCC.* Geneva, Switzerland 151.

Tewolde, H., Fernandez, C.J., Erickson, C.A. (2006). wheat cultivars adapted to post-heading high temperature stress. *J. Agron. Crop Sci., 192*, 111-120.
[http://dx.doi.org/10.1111/j.1439-037X.2006.00189.x]

Thomason, K., Babar, M.A., Erickson, J.E., Mulvaney, M., Beecher, C., MacDonald, G. (2018). Comparative physiological and metabolomics analysis of wheat (*Triticum aestivum* L.) following post-anthesis heat stress. *PLoS One, 13*(6), , e0197919..
[http://dx.doi.org/10.1371/journal.pone.0197919] [PMID: 29897945]

Vijayalakshmi, K., Fritz, A.K., Paulsen, G.M., Bai, G., Pandravada, S., Gill, B.S. (2010). Modeling and mapping QTL for senescence-related traits in winter wheat under high temperature. *Mol. Breed., 26*, 163-175.
[http://dx.doi.org/10.1007/s11032-009-9366-8]

Vishwakarma, H., Junaid, A., Manjhi, J., Singh, G.P., Gaikwad, K., Padaria, J.C. (2018). Heat stress transcripts, differential expression, and profiling of heat stress tolerant gene TaHsp90 in Indian wheat (*Triticum aestivum* L.) cv C306. *PLoS One, 13*(6), , e0198293..
[http://dx.doi.org/10.1371/journal.pone.0198293] [PMID: 29939987]

Waines, J. (1994). High temperature stress in wild wheats and spring wheats. *Funct. Plant Biol., 21*, 705.
[http://dx.doi.org/10.1071/PP9940705]

Wang, J., Qin, Q., Pan, J., Sun, L., Sun, Y., Xue, Y., Song, K. (2019). Transcriptome analysis in roots and leaves of wheat seedlings in response to low-phosphorus stress. *Sci. Rep., 9*(1), 19802.
[http://dx.doi.org/10.1038/s41598-019-56451-6] [PMID: 31875036]

Wang, W., Vinocur, B., Shoseyov, O., Altman, A. (2004). Role of plant heat-shock proteins and molecular chaperones in the abiotic stress response. *Trends Plant Sci., 9*(5), 244-252.
[http://dx.doi.org/10.1016/j.tplants.2004.03.006] [PMID: 15130550]

Wang, X., Hou, L., Lu, Y., Wu, B., Gong, X., Liu, M., Wang, J., Sun, Q., Vierling, E., Xu, S. (2018). Metabolic adaptation of wheat grain contributes to a stable filling rate under heat stress. *J. Exp. Bot., 69*(22), 5531-5545.
[http://dx.doi.org/10.1093/jxb/ery303] [PMID: 30476278]

Wani, S.H. (2019). *Recent Approaches in Omics for Plant Resilience to Climate Change.* Springer.
[http://dx.doi.org/10.1007/978-3-030-21687-0]

Waraich, EA, Ahmad, R, Ashraf, MY (2011). Improving agricultural water use efficiency by nutrient management in crop plants. *Acta Agric. Scand. B Soil Plant Sci., 61*, 291-304.
[http://dx.doi.org/10.1080/09064710.2010.491954]

Woodward, F.I., Lake, J.A., Quick, W.P. (2002). Stomatal development and $CO_2$: ecological consequences. *New Phytol., 153*, 477-484.
[http://dx.doi.org/10.1046/j.0028-646X.2001.00338.x]

Xu, R., Sun, Q., Zhang, S. (1996). Chromosomal location of genes for heat tolerance as measured by membrane thermostability of common wheat cv. Hope. *Hereditas, 18*, 1-3.

Xu, S., Li, J., Zhang, X., Wei, H., Cui, L. (2006). Effects of heat acclimation pretreatment on changes of membrane lipid peroxidation, antioxidant metabolites, and ultrastructure of chloroplasts in two cool-season

turfgrass species under heat stress. *Environ. Exp. Bot., 56*, 274-285.
[http://dx.doi.org/10.1016/j.envexpbot.2005.03.002]

Xu, X.-L., Zhang, Y.-H., Wang, Z-M. (2004). Effect of heat stress during grain filling on phosphoenolpyruvate carboxylase and ribulose-1,5-bisphosphate carboxylase/oxygenase activities of various green organs in winter wheat. *Photosynthetica, 42*, 317-320.
[http://dx.doi.org/10.1023/B:PHOT.0000040608.97976.a3]

Xu, Y., Zhan, C., Huang, B. (2011). Heat shock proteins in association with heat tolerance in grasses. *Int. J. Proteomics, 2011,* , 529648..
[http://dx.doi.org/10.1155/2011/529648] [PMID: 22084689]

Yang, F., Jørgensen, A.D., Li, H., Søndergaard, I., Finnie, C., Svensson, B., Jiang, D., Wollenweber, B., Jacobsen, S. (2011). Implications of high-temperature events and water deficits on protein profiles in wheat (*Triticum aestivum* L. cv. Vinjett) grain. *Proteomics, 11*(9), 1684-1695.
[http://dx.doi.org/10.1002/pmic.201000654] [PMID: 21433286]

Yang, J., Sears, R., Gill, B., Paulsen, G. (2002). Genotypic differences in utilization of assimilate sources during maturation of wheat under chronic heat and heat shock stresses. *Euphytica, 125*, 179-188.
[http://dx.doi.org/10.1023/A:1015882825112]

Yang, J., Zhang, J. (2006). Grain filling of cereals under soil drying. *New Phytol., 169*(2), 223-236.
[http://dx.doi.org/10.1111/j.1469-8137.2005.01597.x] [PMID: 16411926]

Yang, X., Tian, Z., Sun, L., Chen, B., Tubiello, F.N., Xu, Y. (2017). The impacts of increased heat stress events on wheat yield under climate change in China. *Clim. Change, 140*, 605-620.
[http://dx.doi.org/10.1007/s10584-016-1866-z]

Yeh, C.-H., Kaplinsky, N.J., Hu, C., Charng, Y.Y. (2012). Some like it hot, some like it warm: phenotyping to explore thermotolerance diversity. *Plant Sci., 195*, 10-23.
[http://dx.doi.org/10.1016/j.plantsci.2012.06.004] [PMID: 22920995]

Zaharieva, M., Gaulin, E., Havaux, M., Acevedo, E., Monneveux, P. (2001). Drought and heat responses in the wild wheat relative *Aegilops geniculata* roth: Potential interest for wheat improvement. *Crop Sci., 41*, 1321-1329.
[http://dx.doi.org/10.2135/cropsci2001.4141321x]

Zang, X., Geng, X., He, K., Wang, F., Tian, X., Xin, M., Yao, Y., Hu, Z., Ni, Z., Sun, Q., Peng, H. (2018). Overexpression of the wheat (*Triticum aestivum* L.) *TaPEPKR2* gene enhances heat and dehydration tolerance in both wheat and *Arabidopsis. Front. Plant Sci., 9*, 1710.
[http://dx.doi.org/10.3389/fpls.2018.01710] [PMID: 30532762]

Zhang, X., Li, K., Xing, R., Liu, S., Li, P. (2017). Metabolite profiling of wheat seedlings induced by chitosan: revelation of the enhanced carbon and nitrogen metabolism. *Front. Plant Sci., 8*, 2017.
[http://dx.doi.org/10.3389/fpls.2017.02017] [PMID: 29234335]

Zhang, Y., Lou, H., Guo, D., Zhang, R., Su, M., Hou, Z., Zhou, H., Liang, R., Xie, C., You, M., Li, B. (2018). Identifying changes in the wheat kernel proteome under heat stress using iTRAQ. *Crop J., 6*, 600-610.
[http://dx.doi.org/10.1016/j.cj.2018.04.003]

# CHAPTER 2

# Genetic Enhancement of Heat Tolerance in Maize Through Conventional and Modern Strategies

**M. G. Mallikarjuna[1,*], Jayant S. Bhat[2], Firoz Hossain[1], Palanisamy Veeraya[1], Akshita Tyagi[1], Chikkappa G. Karjagi[3] and H. C. Lohithaswa[4]**

[1] *Division of Genetics, ICAR-Indian Agricultural Research Institute, New Delhi-110012, India*

[2] *Regional Centre, ICAR-Indian Agricultural Research Institute, Dharwad-580005, India*

[3] *Unit Office, ICAR-Indian Institute of Maize Research, New Delhi-110012, India*

[4] *Department of Genetics and Plant Breeding, College of Agriculture (University of Agricultural Sciences, Bengaluru), Mandya-571405, India*

**Abstract:** Globally, maize is an important crop and serves as a livelihood for millions of marginal farmers across South Asia and sub-Saharan Africa. However, heat stress has become a globally prominent and growing concern for maize farmers, owing to its adverse impact on maize growth and productivity. In addition, the mean maximum temperature may increase by 2.1–2.6°C in 2050, with significant temporal and spatial variations, across South Asia. Further, the heat-stressed areas would increase to 21% from the current baseline. Heat stress is known to induce a series of morpho-physiological, anatomical and molecular changes in maize, thereby affecting growth and development, which ultimately leads to a drastic reduction in the grain yield. Regulation of osmoprotectants, detoxification of excess reactive oxygen species (ROS), expression of heat-responsive/shock genes, and change of plant phenology help in the development of heat tolerance. The molecular basis of heat stress tolerance mechanisms has been appreciably understood and updated using the innovative physiological and molecular tools. Further, functional genomics strategies resulted in the identification of genes and regulatory pathways involved in heat stress tolerance in maize. Several attempts have been made in breeding heat-tolerant maize cultivars. The availability of genomic resources, accessibility to sequence information and millions of SNP markers in maize facilitated the selection for heat tolerance at the genome-level. Genomics-assisted mapping revealed several QTL and interactions for heat stress functional adaptive traits. The new breeding approaches like doubled haploid inducers, genome editing tools and high throughput phenomics at the breeders' disposal are opening up a new era in maize breeding for development of heat resilient maize hybrids.

* **Corresponding author M. G. Mallikarjuna:** Division of Genetics, ICAR-Indian Agricultural Research Institute, New Delhi-110012, India; E-mails: MG.Mallikarjuna@icar.gov.in; mgrpatil@gmail.com

**Keywords:** Climate Change, Functional Genomics, Genomics-Assisted Breeding, Heat, Maize, Resilience.

## INTRODUCTION

Anthropogenic climate change is rapidly changing the cropping pattern, and agronomic practices developed over thousands of years due to changing abiotic and biotic stress dynamics. During this climate change era, global warming is becoming a major concern to ensure sustainable food production (Ainsworth & Ort, 2010). An increase in mean annual temperature beyond the optimum level also results in greater yield penalty in crops, which in turn affect food and feed availability (Tigchelaar, Battisti, Naylor, & Ray, 2018). The projections estimated 31-50% losses of grain yield in major food crops *viz.*, wheat (Balla *et al.*, 2011), rice (Peng *et al.*, 2004), maize (Bassu *et al.*, 2014), soybean (Schlenker & Roberts, 2009), Brassica (Angadi *et al.*, 2000); sorghum (Tack, Lingenfelser, & Jagadish, 2017), and groundnut (Cooper *et al.*, 2009). Globally, maize is the widely grown crop and leading the cereals in terms of production. In the developing countries of Africa, Asia and Latin America, maize is one of the major contributors towards food and nutritional security (Agrawal, Mallikarjuna, & Gupta, 2018; Mallikarjuna *et al.*, 2014; Prasanna, 2012). The percent share of maize as a source of food ranges from 61% in the Mesoamerica region to 4% in South Asia (Shiferaw, Prasanna, Hellin, & Bänziger, 2011). Along with rice and wheat, maize provides at least 30% of the food calories to more than 4.5 billion people in 94 developing countries, including 900 million poor consumers (Shiferaw *et al.*, 2011). In addition, maize also has a diversified usage *viz.*, feed for poultry and livestock, bioenergy production and raw material for various industries (Agrawal *et al.*, 2018; Cairns *et al.*, 2012a; Mallikarjuna *et al.*, 2014, 2015; Prasanna, 2012).

Presently, maize growing area is showing an annual growth rate of 2.7%, 3.1%, and 4.6% in Africa, Asia and Latin America, respectively (http://www.fao.org/faostat). Further, the demand for maize in the developing world will double by 2050 to meet the requirement of ~10 billion people (Mickelbart, Hasegawa, & Bailey-Serres, 2015; Rosegrant, Ringler, & Zhu, 2009). However, maize is affected by various abiotic and biotic production constraints *viz.*, drought, heat, waterlogging, pests and diseases. Among the various stresses, heat is difficult to manage through cultural operations, unlike pests and diseases. Heat stress is becoming a regular phenomenon owing to global warming and changing climate. In rainfed maize, each degree spent above 30°C reduces the final yield by 1% and 1.7% per day under optimum irrigation and drought conditions, respectively (Lobell, Bänziger, Magorokosho, & Vivek, 2011). Based on the growth stage, the overall extent of heat-induced grain yield losses in maize goes up to >50% (Bassu *et al.*, 2014). Recently, (Tigchelaar *et al.* 2018) predicted the maize yield reduction in response to 2°C and 4°C increase in mean temperature. Hence, to meet the maize requirement of the growing population, it is imperative to enhance the maize yield sustainably in the era of climate change and global warming. Among the various approaches available, genetic manipulation of maize could be the most sustainable and economical approach to address the adverse effects of heat stress on maize. Genetic and molecular mechanisms play a crucial role in assigning of heat tolerance and survival of plants. Therefore, the successful development of heat-tolerant maize hybrids necessitates the understanding of genetics, physiological and molecular mechanisms associated with heat tolerance. In this chapter, we have summarised the updates on the genetic and molecular basis and breeding approaches of heat tolerance in maize.

## THE RESPONSES OF MAIZE TO HEAT STRESS

The sensitivity of maize to heat stress varies with crop duration, plant architecture, genotypes, and degree and intensity of heat stress. Heat stress affects the plants by inducing a cascade of morpho-physiological and molecular changes (Fahad *et al.*, 2017). The important phenological and physio-molecular responses are briefly discussed under the following broad headings (Fig. **1**).

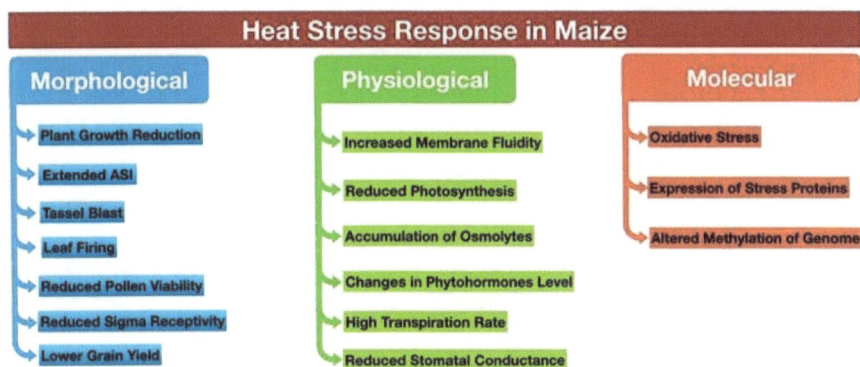

**Fig. (1).** Major morphological, physiological and molecular effects of heat stress on maize.

## Morphological Responses

The heat stress is known to affect the various vegetative and reproductive traits, which are directly and indirectly associated with grain yield. The exposure of the maize seeds to 45°C reduced the germination percentage and shoot dry mass from 67 to 55%, and 6.13 to 4.16g, respectively (Ashraf & Hafeez, 2004). One of the striking morphological vegetative responses to heat stress is leaf firing, which mainly occurs when heat stress associated with low relative humidity (Zaidi *et al.*, 2016). In addition to higher day temperature, increased night temperature causes decreased maize growth and development. Higher night temperature, coupled with narrow diurnal temperature amplitude, resulted in a negative impact on vegetative growth in maize (Sunoj, Shroyer, Jagadish, & Prasad, 2016).

The proper seed set in maize is the result of various reproductive mechanisms *viz.*, male and female organs differentiation and development, anthesis, silking, anthesis-silking interval (ASI), pollen shedding, silk emergence, pollen germination and pollen tube elongation on the stigma, fertilisation, zygote development and grain filling duration (Dresselhaus & Franklin-Tong, 2013; Mayer, Rattalino Edreira, & Maddonni, 2014; Schoper, Lambert, & Vasilas, 1987). These reproductive stages and processes of maize are the most sensitive to heat stress as compared to any other vegetative stage. Maximum exposure of tassels to heat stress enhances the probability of heat-induced pollen damage (Cairns *et al.*, 2012a, 2012b). The heat stress in maize results in tassel blast characterised by drying of the maximum and complete portion of tassel without pollen extrusion (Zaidi *et al.*, 2016). The synchronisation between anthesis and silking is crucial for proper fertilisation and seed set. The occurrence of heat stress during anthesis and silking stages results in more delayed silking than anthesis, which expands the ASI (Cicchino, Rattalino Edreira, Uribelarrea, & Otegui, 2010). Further, increased temperature results in pollen desiccation, lower pollen viability, reduced pollen number and shedding and increased silk death (Dupuis & Dumas, 1990; Pingali, 2001; Schoper *et al.*, 1987).

In addition to flowering traits, grain-filling rate and kernel setting are other most extensively studied reproductive traits in maize under heat stress. High-temperature stress shortens the grain filling duration resulting in the lower kernel test weight (Mayer *et al.*, 2014; Tao *et al.*, 2016). The direct post-silking exposure of maize from the day/night temperature of 30/15°C to 35/25°C for 18 days resulted in a 16.9−44.4% reduction in the grain weight per plant (Pingali, 2001). Similarly, a linear decrease in the matured kernel weight was observed with temperature exceeding 22°C (Singletary, Banisadr, & Keeling, 1994). The kernel weight decreased by 7% was found under a day/night temperature stress of 33.5/25°C (Wilhelm, Mullen, Keeling, & Singletary, 1999). Similarly, in spring maize, the occurrence of temperature ≥33°C for 6−17 days during the grain filling period negatively affected the 1000-kernel weight and kernel number (Tao *et al.*, 2013). The kernel set percentage also indicates the sigma receptivity during the pollination period. Heat stress-induced kernel set reduction of 57−80% was observed in maize (Alam *et al.*, 2017).

## Physiological Responses

Heat stress in maize displays a series of changes in membrane stability, photosynthesis, respiration rate, osmolytes concentration, hormone levels and various other physiological functions (Tiwari & Yadav, 2019; Wahid, Gelani, Ashraf, & Foolad, 2007). Disturbances in the physiological processes affect the plant survivability and economic grain yield. The effects of heat stress on maize physiology are described under the following headings.

### *Photosynthesis*

Photosynthesis is the most important physiological process and very sensitive to heat stress (Berry & Bjorkman, 1980; Crafts-Brandner & Salvucci, 2002). In addition to the disturbance in photosynthetic machinery, heat stress also affects the photosynthates distribution, which subsequently results in reduced grain yield and biomass (Rattalino Edreira & Otegui, 2012; Reed & Singletary, 1989; Wahid *et al.*, 2007). Significant and positive correlations were identified between photosynthetic parameters and grain yield, indicating the importance of photosynthetic machinery in maize grain yield under heat stress (Galic *et al.*, 2019).

Chloroplasts are the structural and functional units of photosynthesis and comprise thylakoids and a matrix. Thylakoid membrane holds photosynthetic electron transport, phosphorylation and light-harvesting machineries *viz.*, photosystem I (PSI) and photosystem II (PSII). Heat stress increases membrane permeability resulting in the release of electrolytes, loss of grana stacking and swelling of grana (Ashraf & Hafeez, 2004; Chen & Burris, 1991; Rodríguez, Canales, & Borras-Hidalgo, 2005; Savchenko, Klyuchareva, Abramchik, & Serdyuchenko, 2002; Wang, Chen, He, & Guo, 2018; Yang, Rhodes, & Joly, 1996). Moreover, heat stress affect the chlorophyll biosynthesis and results in decreased photosynthates production (Hasanuzzaman, Nahar, Alam, Roychowdhury, & Fujita, 2013). High temperature significantly reduced the chlorophyll content and photosynthetic activity. Consequently, a reduction of 50-60% in net photosynthesis rate was observed (Ben-Asher, Garcia Y Garcia, & Hoogenboom, 2008; Caers, Rudelsheim, Onckelen, & Horemans, 1985). Further, the reduction of total chlorophyll components *viz.*, chlorophyll-a and chlorophyll-b were reported in response to heat stress (Dinler, Demir, & Kompe, 2014; Zhu, Song, Liu, & Liu, 2011). The photosystem II (PSII) activity is largely reduced or even ceased under heat stress even ceased under heat stress (Rodríguez *et al.*, 2005; Tao *et al.*, 2016). The decrease in the quantum yield of PSII electron transport ($\Phi$PSII), PSII photochemistry (Fv'/Fm'), photochemical quenching coefficient (qP), and a massive increase in non-photochemical quenching coefficient (qN) were observed under heat stress in maize (Lu & Zhang, 2000; Zhu *et al.*, 2011).

In addition to direct effects on the photosynthetic machinery, heat stress also reduces the photosynthetic carbon assimilation *via* temperature-induced reduction of pyruvate carboxylase enol phosphate (PEPC) and Rubisco activity. The rapid increase of the leaf temperature to 45°C led to nearly complete inactivation of Rubisco activity in maize (Crafts-Brandner & Salvucci, 2002). The heat-tolerant maize lines showed the higher activity of PEPCase and RuBPCase activities in leaf as compared to sensitive genotypes at a daily mean temperature of ≥35°C (L. Zhao, Li, Liu, Wang, & Seng, 2012). Recent findings demonstrated that nine of thirty-five key metabolites largely explain the variance of photosynthetic efficiency in response to heat stress recovery. Further, the increase in PEPC activity was reported during recovery from heat stress in both elevated and ambient $CO_2$ conditions (Qu, Chen, Bunce, Zhu, & Sicher, 2018).

## *Water Relations and Osmotic Adjustments*

Sufficient tissue moisture content is necessary to stabilise the metabolic process in plants. The optimum soil moisture content increases the leaf moisture content under high temperature and enhances heat tolerance in maize (Tiwari & Yadav, 2019; Zhu *et al.*, 2011). However, in the era of climate change, heat stress commonly coincides with water stress, especially in tropical and subtropical maize growing areas (Cairns *et al.*, 2012; Tesfaye *et al.*, 2018). The effects of heat stress are more destructive under drought conditions (Cairns *et al.*, 2013; Machado & Paulsen, 2001). Heat stress enhances the transpiration rate to maintain canopy temperature. When heat stress is coupled with drought, plants will not be able to open their stomata, and their leaf temperature increases rapidly (Garcia y Garcia, Abritta, Soler, & Green, 2014). Heat stress in maize decreases the leaf tissue water content and transpiration rate. However, the combined heat and drought stress decreased the transpiration rate owing to lower stomatal conductance (Hussain *et al.*, 2019). In addition to the direct effect of heat stress on vapour pressure deficit (VPD) and transpiration rate, it also changes the plant hydraulic conductance and water supply to the leaf surface from roots. Further, the sensitivity of aquaporin transporters to temperature partially determines the plant's internal hydraulic conductivity under heat stress environments (Yang *et al.*, 2012). Plants, including maize accumulate several organic compounds of low molecular mass in response to heat stress such as osmolytes, sugars and sugar alcohols (polyols), proline, ammonium and sulphonium compounds (Wahid *et al.*, 2007). Osmolytes accumulation during stress enhances the stress tolerance in crops. The accumulation of endogenous betaine in maize enhances heat tolerance in maize (Li & Zhu, 2015). The exposure of maize seedlings to 45±2°C for 2h resulted in increased accumulation of proline, glycine, betaine, spermidine, putrescine, spermine and total soluble sugars (Mohamed, Ashry, & Ghonaim, 2019).

## Oxidative Stress

Oxidative stress occurs as secondary stress owing to enhanced production of reactive oxygen species (ROS) in various metabolic pathways under heat stress. The sensitivity of metabolic pathways towards heat stress regulates the production of ROS (Asada, 2006; Qu, Ding, Jiang, & Zhu, 2013). The most commonly produced ROS in plants are singlet oxygen ($^1O_2$), superoxide radical ($O_2^{\cdot-}$), hydrogen peroxide ($H_2O_2$) and hydroxyl radical ($OH^{\cdot}$) (Asada, 2006; Hasanuzzaman *et al.*, 2013). The ROS reacts with the majority of biomolecules in the cells *viz.*, pigments, proteins, lipids and nucleic acids, and severely affect the plant's growth and development during heat stress (Xu, Li, Zhang, Wei, & Cui, 2006). In addition to the direct quantification of ROS, malondialdehyde (MDA) and hydrogen peroxide contents are used as the measure of oxidative stress in plants, since MDA provides the measure of lipid peroxidation (Kumar, Gupta, & Nayyar, 2012). The maize hybrids subjected to heat stress (38 °C/30 °C) showed significantly increased $O_2^{\cdot-}$, $H_2O_2$, and MDA production, although their levels are lower in the tolerant genotype as compared to sensitive (Hussain *et al.*, 2019). A similar trend of $H_2O_2$ accumulation under heat stress was reported in sensitive and tolerant maize genotypes (Gong, Chen, Li, & Guo, 2001). Additionally, the studies revealed the negative association between the increased concentration of ROS and key metabolic activities in plants (Sejima, Takagi, Fukayama, Makino, & Miyake, 2014).

## Hormonal Response

Hormones are the key compounds known to modulate the various developmental and physiological responses during heat stress. Among various phytohormones, abscisic acid (ABA), auxin and cytokinin play major roles in photosynthate remobilisation and grain filling (Sehgal *et al.*, 2018; Yu, Lo, & Ho, 2015). Synthesis of auxin-binding protein (ABP1) strongly declined in response to heat stress in maize (Oliver, Venis, Freedman, & Napier, 1995). Heat stress shifts the hormonal balance of developing kernels in maize. Exposing of maize plants at the reproductive stage revealed a non-significant correlation between ABA and cytokinin levels as against significant negative correlation in non-stressed kernels (Cheikh & Jones, 1994). Further, Caers *et al.* (1985) reported the association between phytohormone content and chloroplast biogenesis in maize. The primary leaves in maize seedlings showed low cytokinin concentration under heat stress as compared to the below detectable level in etiolated leaves under heat stress. Heat stress further reduced salicylic acid (SA), increased ABA and indole acetic acid (IAA) levels in susceptible maize genotypes against increased SA and IAA concentration in tolerant genotype (Dinler *et al.*, 2014).

## Molecular Responses

### Heat Shock Proteins (HSPs)

HSPs play important roles in the heat stress responses and thermotolerance in plants. Synthesis and accumulation of HSPs were established during heat stress in all the living organisms, including plants. In plants, based on molecular weight, HSPs are classified into five different families *viz.*, HSP100, HSP90, HSP70, HSP60 and HSP 20 (small HSP) (Sanmiya, Suzuki, Egawa, & Shono, 2004; Swindell, Huebner, & Weber, 2007). The HSPs act as molecular chaperones by inhibiting irreversible aggregation of other proteins and help in refolding of proteins during high-temperature stress to regulate the cellular homeostasis (Swindell *et al.*, 2007; Tripp, Mishra, & Scharf, 2009). Additionally, HSPs prevent the heat stress-induced denaturation of other proteins and help in the sorting of proteins in the cell (Pegoraro, Mertz, da Maia, Rombaldi, & de Oliveira, 2011; Swindell *et al.*, 2007; Tripp *et al.*, 2009). Heat stress in maize results in enhanced accumulation of HSPs in tissues such as germinating seeds and embryo (Nieto-Sotelo *et al.*, 2002), pollen (Young *et al.*, 2001), and ear (Kandel, Ghimire, & Shrestha, 2018). The heat shock factors (HSF) serve as the terminal component of signal transduction and regulate the expression of genes encoding HSPs (Guo *et al.*, 2016; Hasanuzzaman *et al.*, 2013). In addition to the regulatory role, HSFs also serve as sensors during heat stress (Davletova *et al.*, 2005). The heat shock factors HsfA1 and HsfA2 are the major heat stress response factors regulating the heat stress response in plants (Guo *et al.*, 2016; Morimoto, 1998).

### Signalling Transduction of Heat Stress

The heat stress signal perception and transduction are critical for plants to assign heat stress tolerance. Various signalling pathways and molecules are involved in sensing heat stress leading to the generation of the adaptive responses. Membrane systems play a very important role in the sensing of heat stress in plants. The lipid composition of membranes changes with respect to the surrounding temperature, and can thus modulate heat-sensing components present in and within the membrane system. Enhanced permeability of membranes triggers a specific transient $Ca^{2+}$ influx across the plasma membrane (Saidi, Finka, & Goloubinoff, 2011). The intensity of the $Ca^{2+}$ signature and of the following heat stress response was reduced in the tissues with more saturated lipids and thus more rigid membranes (Saidi *et al.*, 2010). The reports showed that treatment with membrane fluidizer benzyl alcohol resulted in a similar cascade of events of heat tolerance through enhanced $Ca^{2+}$ influx within minutes, HSP expression within hours and the development of thermotolerance within days (Saidi *et al.*, 2005, 2011, 2009; Suri & Dhindsa, 2008). On the other hand, exposure of plant membranes to rigidifying agent dimethylsulfoxide, reduces the heat-mediated expression of HSPs (Suri & Dhindsa, 2008). These findings supported the existence of membrane-associated thermosensors that can perceive the heat signals and respond appropriately. Heat-induced membrane fluidity results in two important signalling pathways during heat stress *viz.*, 1) lipid signalling, and 2) $Ca^{2+}$ signalling pathways.

In lipid signalling, heat stress results in the activation of phospholipase D (PLD) and phosphatidylinositol-4-phosphate 5-kinase (PIPK), leading to the accumulation of lipid signalling molecules *viz.*, phosphatidic acid, phosphatidylinositol-4,5-bisphosphate (PIP2) and D-myoinositol-1,4-5-trisphosphate (IP3) (Mishkind, Vermeer, Darwish, & Munnik, 2009). Additionally, the reduced phospholipase C9 activity correlated with reduced IP3 concentration reduced thermotolerance and downregulation of sHSPs indicating the potential role of phospholipase in heat stress signalling (Zheng *et al.*, 2012).

In $Ca^{2+}$ signalling pathway, increased membrane fluidity during high temperature opens the specific calcium channel and facilitates the inward flux of calcium ions ($Ca^{2+}$) (Mittal *et al.*, 2017; Mittler, Finka, & Goloubinoff, 2012). The increased $Ca^{2+}$ regulates multiple heat-responsive signalling pathways in plants (Goraya *et al.*, 2017; Mittler *et al.*, 2012; X. Wang, Xu, Cai, Wang, & Dai, 2017). In the cell, sensing of altered $Ca^{2+}$ concentration by calcium-binding proteins or calcium sensors (calmodulin, calmodulin-like proteins, calcineurin B- like proteins, and calcium-dependent protein kinases) results in the downstream expression events (Mittal *et al.*, 2017; Sanders, Brownlee, & Harper, 1999). In addition to $Ca^{2+}$ binding proteins/sensors, increased $Ca^{2+}$ concentration activates the ROS-producing enzyme NADPH oxidase (Suzuki, Sejima, Tam, Schlauch, & Mittler, 2011). During heat, calmodulin CaM3 is known to activate the transcription factors WRKY39, MYB and HSFs (Li, Zhou, Chen, Huang, & Yu, 2010; Liu, Liao, & Charng, 2011; X. Wang *et al.*, 2017). Further, activation of calcium/calmodulin-binding protein kinase (CBK) by CaM3 results in the phosphorylation of heat shock transcription factor HSFA1a (Liu *et al.*, 2008). Dephosphorylation of HSFA1a is carried by phosphatase PP7 (Reddy, Ali,

Celesnik, & Day, 2011). In addition to CBK, an HSP90/FKBP-dependent kinase, ROF1 (FKBP62) is also known to phosphorylate HSFs (Meiri & Breiman, 2009). Phosphorylation of multiprotein bridging factor 1c (MBF1c) functions as a transcriptional regulator of 36 different transcripts during heat stress, including DRE-binding protein 2A (DREB2A) and two heat shock transcription factors (HSFs) (Bokszczanin & Fragkostefanakis, 2013; Suzuki, Sejima, *et al.*, 2011). The pictorial representation of heat stress-induced $Ca^{2+}$ dependent signal transduction is given in Fig. (**2**).

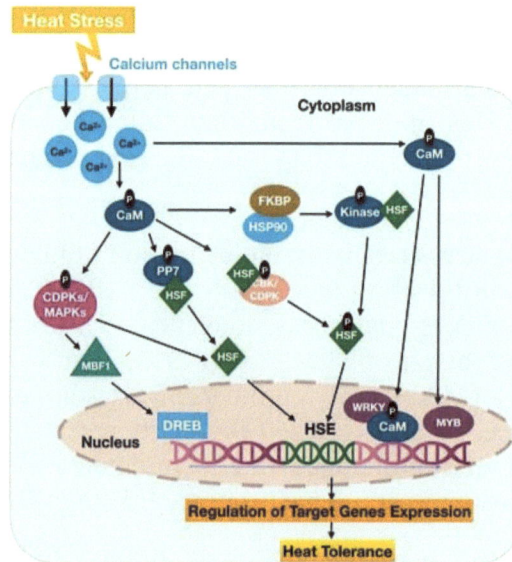

**Fig. (2).** Schematic representation of $Ca^{2+}$ signal transduction pathways in response to heat stress in plants (Modified and adopted from Mittler *et al.*). Heat stress affects membrane stability and activates a PM calcium channel, resulting in an inward flux of calcium. In cytoplasm $Ca^{2+}$ ions binds the calmodulin (CaM3) and acti]ates various kinases and transcriptional regulators of the heat stress response, *viz.*, HSFs, MBF1c, WRKY and DREB. Subsequently, the activated heat responsive transcriptional regulators binds to the heat-shock elements (HSE) located in the promoter region of HSP genes and results enhanced heat stress tolerance.

## ADAPTIVE MECHANISMS FOR HEAT TOLERANCE IN MAIZE

Heat tolerance is the ability of the plants to sustain the economic yield through various heat stress adaptive mechanisms. Plants have evolved a variety of adaptive mechanisms *viz.*, early maturation, leaf orientation, transpirational cooling and changes in membrane lipid composition, increased antioxidants and heat shock protein activity (Hasanuzzaman *et al.*, 2013; Rodríguez *et al.*, 2005; Wahid *et al.*, 2007). Maize genotypes showed variable morpho-physiological responses to heat stress. In maize, stigma-receptivity is less sensitive as compared to pollen viability; however, delayed stigma initiation under heat stress prolongs the ASI beyond 2-4 days. Therefore, the tolerant genotypes possess narrow ASI and pollen viability under heat stress (Alam *et al.*, 2017). Leaf firing and tassel blast are the visible key heat stress symptoms in maize. The canopy temperature during daytime is more than the surrounding air temperature. Therefore, maintaining of leaf temperature to sustain the metabolic activities is imperative to switch on the transpiration cooling of leaves through increasing the transpiration rate. Subsequently, to balance the enhanced transpiration rate during heat stress, tolerant maize germplasm is expected to increase the root mass for efficient water foraging from soil (Khanna, Kaur, & Gupta, 2017).

Plants evolved antioxidant defence mechanisms to counteract the damages caused by ROS. The antioxidant defence mechanisms operate through enzymatic and non-enzymatic components. The enzymatic part includes superoxide dismutase (SOD), catalase (CAT), guaiacol peroxidase (GPX), ascorbate peroxidase (APX), and glutathione reductase (GR) whereas, non-enzymatic components include comprises of flavonoids, anthocyanin, carotenoids and ascorbic acid (AA) *etc.* (Asthir, 2015; Suzuki, Miller, *et al.*, 2011). During heat stress, tolerant maize genotypes showed higher production of antioxidative enzymes in contrast to susceptible genotypes to overcome the oxidative damages. Hussain *et al.* Hussain *et al.* (2019) reported high activities of SOD, POD, CAT in tolerant as against susceptible genotype. In addition to antioxidants, increased concentration of acetylcholinesterase in endodermal cells is related to heat stress tolerance in maize seedlings (Yamamoto, Sakamoto, & Momonoki, 2011).

The heat stress further stimulates the expression and accumulation of HSPs and other proteins to induce tolerance mechanisms. Heat stress increases HSPs content, which subsequently assembles into heat shock granules (HSGs) to protect the protein synthesis machinery from protein denaturation (Miroshnichenko *et al.*, 2005). Small HSPs (sHSPs) played a vital role in the protection of PS II during heat stress and showed heat-responsive accumulation in maize leaves (Hu *et al.*, 2010). The reports demonstrated the strong interaction between maize sHSP26 and ATP synthase subunit β, chlorophyll a–b binding protein, oxygen-evolving enhancer protein-1 and photosystem I reaction centre subunit IV protect the maize chloroplasts (Hu, Yang, *et al.*, 2015). Similarly, chloroplast protein synthesis elongation factor (EF-Tu) is known to assign heat stress tolerance in maize (Bhadula *et al.*, 2001; Ristic *et al.*, 2004). In chloroplast, EF-Tu protein acts as a molecular chaperone and protects chloroplast proteins from thermal aggregation and inactivation (Rao, Momcilovic, Kobayashi, Callegari, & Ristic, 2004). Bhadula *et al.* (2001) and Momcilovic & Ristic, (2004) showed that the heat-induced enhanced accumulation of EF-Tu is specific to heat-tolerant maize genotype ZPBL1304, as compared to senitive genotype ZPL389 (Bhadula *et al.*,

2001; Momcilovic & Ristic, 2004). Momcilovic and Ristic (2007) demonstrated the accumulation of both EF-Tu transcript and EF-Tu protein in the heat-tolerant genotype as compared to sensitive genotype.

## FUNCTIONAL GENOMICS

The heat stress response in plants involves several pathways, genes, cellular compartments, regulatory elements and networks. Plants have evolved various functional adaptive strategies at the cellular and molecular levels to cope with the heat stress *viz.*, alteration of transcriptional control and signalling cascades, enhanced production of antioxidants, osmoprotectants and stress proteins (Almeselmani, Deshmukh, Sairam, Kushwaha, & Singh, 2006; Kotak *et al.*, 2007; Mittler *et al.*, 2012; Qian, Ren, Zhang, & Chen, 2019). Advances in the genomics allowed monitoring of changes in genomic snapshots in response to stress tolerance, which in turn facilitates the dissection of systems biology operating behind the adaptive mechanisms (Arora *et al.*, 2017; Thirunavukkarasu *et al.*, 2017).

Accessibility of genomic regions for the expression machinery is a fundamental requirement for the expression of genes. The epigenetic regulation of the genomic regions may contribute to the transcriptomic and phenotypic differences (Eichten & Springer, 2015; Noshay *et al.*, 2019). Stable inheritance of DNA methylation through mitosis or meiosis in maize for various traits provides the epigenetic basis for trait inheritance (Noshay, Crisp, & Springer, 2018; Yong, Hsu, & Chen, 2016). Appreciable efforts were made to investigate the maize epigenome, although very few were reported for heat stress (Eichten & Springer, 2015; Noshay *et al.*, 2018; Qian, Hu, Liao, Zhang, & Ren, 2019). Profiling of DNA methylation in response to heat stress in the B73 inbred line showed 325 differentially methylated genes mainly enriched in pathways associated with spliceosome, homologous recombination, RNA transport, ubiquitin-mediated proteolysis and carbon metabolism (Qian *et al.*, 2019).

Genome-wide expression profiling at the seedling stage in response to heat stress was conducted in European dent and flint maize inbred lines with distinctive responses to heat stress. The study identified 607 heat-responsive genes and 39 heat tolerance genes. Among the heat-responsive genes, 14 genes falling under HSPs, extracellular ribonuclease activity, peptides transport protein, signal transduction, nicotianamine synthases, *etc.* were expressed across the inbred lines (Frey, Urbany, Hüttel, Reinhardt, & Stich, 2015). Comparative transcriptome profiling of heat- resistant (XT) and heat-sensitive (ZF) sweet maize varieties under heat stress commonly upregulated 516 and down-regulated 1,261 genes across the treatments *viz.*, XT/ZF, XT0.5/ZF0.5 and XT3/ZF3. However, the maximum number of differentially expressed genes [DEGs] was observed in treatment with 3h exposure to heat stress. The KEGG [Kyoto Encyclopedia of Genes and Genomes] found that pathways enriched in DEGS were secondary metabolite biosynthetic pathways and ribosomes. Additionally, the study predicted that the reduced biosynthesis of zeatin and brassinosteroids were associated with enhanced heat tolerance in heat-tolerant sweet maize genotypes (Jiang Shi, Yan, Lou, Ma, & Ruan, 2017). Genome-wide transcriptome assay was employed to elucidate the functional response of maize to various abiotic stresses such as drought, salinity, heat, and cold stresses in the genetic background of B73. A total of 1,661, 2,019, 2,346, and 1,841 DEGs were identified in comparison to the control with salinity, drought, heat, and cold stress, respectively. A total of 167 common genes, including 12 TFs indicated the presence of a common regulatory mechanism(s) in modulating the response targeted abiotic stresses (P. Li *et al.*, 2017). Transcriptome analysis in inbred line B73 under heat stress at the seedling stage resulted in 1029 up- and 828 down-regulated DEGs. In addition to regular genes, the DEGs also encompass the putative TFs belonging to MYB, AP2-EREBP, b-ZIP, bHLH, NAC and WRKY families. This also supported the previous finding on the regulatory role of MYB TFs in assigning heat tolerance (Qian *et al.*, 2019). The KEGG pathway enrichment of DEGs showed the pivotal role of protein processing in endoplasmic reticulum pathways in the regulation of heat stress response. Therefore, the study emphasized the role of heat stress and protein metabolism in maize (Qian, *et al.*, 2019). The genome-wide expression analysis during highly heat-sensitive tetrad stage of pollen development in maize demonstrated the alterations in primary metabolic pathways as a basis for the subsequent production of defective pollens germination under heat stress (Begcy *et al.*, 2019). Further, the study revealed a putative association of nine TF binding sites *viz.*, ABRE, CT-rich, E2F, Em1d, G-box, GA-5, Myb2, PII and Sph with heat tolerance mechanism in maize (Begcy *et al.*, 2019).

A few researchers investigated the proteome and metabolome changes under heat stress in maize. Characterisation of small heat shock proteins in response to combined drought and heat stress identified seven proteins associated with stress *viz.*, cytochrome b6-f complex iron-sulfur subunit, sHSP17.4, sHSP17.2, sHSP26, guanine nucleotide-binding protein b-subunit-like protein, putative uncharacterized protein, and granule-bound starch synthase IIa (Hu *et al.*, 2010). Similarly, phosphoproteome analysis in response to heat stress revealed the differential expression of 75 phosphoproteins (Hu *et al.*, 2015). Additionally, total proteome changes under heat stress detected 61 heat stress-specific proteins falling under the various categories *viz.*, HSP, chloroplast function, signalling, *etc*. (Zhao *et al.*, 2016). The changes in the metabolite profiling of maize were observed in response to various stresses, including heat (Sun *et al.*, 2016a; 2016b). Metabolite profiling of maize leaves in heat and combined heat and drought stresses have given insights on stress-dependent metabolite accumulation. Reduced accumulation of phenylalanine, alanine, 4-aminobutanoate, threonate, xyloside, galactinol, isoleucine, glycerol, malate, glycerate, and phosphate whereas, increased accumulation was witnessed for the metabolites tryptophan, serine, threonine, *beta*-alanine, proline, glutamic acid, pyroglutamic acid, raffinose, myoinositol, succinate, urea, maltose, erythritol, maltitol, and trehalose under heat stress (Obata *et al*., 2015).

## MAIZE BREEDING FOR HEAT TOLERANCE

### Screening Methodologies and Heat Tolerance Indices

The air temperature around the canopy determines the rate of plant desiccation under heat stress. In addition to air temperature, relative humidity (RH) *via* vapour pressure deficit [VPD] contributes towards the extent of damages in maize. Therefore, the simultaneous occurrence of high temperature ($T_{max}$>33°C and $T_{min}$>23°C) with low RH (<40%) is very detrimental to maize growth and development (Zaidi *et al*., 2016). The stringent and reliable screening protocols for heat stress tolerance in maize necessitate the well-defined temperature, RH, screening sites and type of germplasm.

For field-based screening, the screening site(s) and season need to be selected based on climatological data. Heat stress in the targeted sites can be applied by altering the sowing time to ensure coincidence of reproductive growth stages, *i.e*., from tassel emergence to early grain-filling with peak heat stress. Being a complex mechanism, the measurement of heat tolerance in maize requires better tolerance indices for lab-based and field-based screening methods. The traits *viz*., ASI, leaf firing, tassel blast, tassel sterility and grain yield are important indices for heat stress tolerance in the field condition at reproductive stages (Akula *et al*., 2016; Alam *et al*., 2017; Cairns *et al*., 2013; Chen, Xu, Velten, Xin, & Stout, 2012; Lizaso *et al*., 2018; Noor *et al*., 2019). For heat stress experiments, simulation of an appropriate heat stress regime is an utmost important step. The studies conducted using tropical and subtropical maize germplasm in the different locations of India (Hyderabad, Raichur, Bhemarayangudi), Mexico, Zimbabwe and Thailand ensured approximate temperature of $T_{max}$>35°C (day) and $T_{min}$>25°C (night) during reproductive stages (Alam *et al*., 2017; Cairns *et al*., 2013; Noor *et al*., 2019). Similarly, Chen *et al*. (2012) and Rattalino Edreira and Otegui, (2012) tested the maize genotypes for heat stress during the reproductive stage with an average day temperature of $T_{max}$>35°C and >33°C, respectively. On the other hand, stringent high temperature (more than 40°C) was used in the screening of maize flowering and grain filling stages (Yousaf *et al*., 2018).

As compared to field-based screening at reproductive stages, seedling level screening methodologies for heat stress tolerance are time and resource-efficient. However, screening methods at the seedling stage do not provide direct information on the impact of heat stress on grain yield, which is the most important target trait for any maize breeders. The seedling level indices for heat stress tolerance in maize, include growth rate, membrane thermostability, chlorophyll content, leaf temperature, and seedling phenology traits (Frey *et al.*, 2015; Tandzi, Bradley, & Mutengwa, 2019). Various protocols have been reported on the screening of maize for heat tolerance at the seedling stage. The maize seedlings were exposed to mild and severe heat stress of 32°C (day) and 27°C (night), and 38°C (day) and 33°C (night), respectively to investigate the seedling response to heat stress in the dent and flint maize germplasm (Frey *et al.*, 2015). Heat stress was applied on maize seedlings with 40°C day and 25°C night temperature and with 60% RH for 3 days/nights with a gradual increase in temperature of 5°C increments per hour from 25-40°C during the stress period (Tandzi *et al.*, 2019).

Though heat stress occurs more often during summer or spring season in Asia, heat stress also occurs during the rainy season, especially under low moisture stress conditions. Further, apart from the identification of genotypes for exclusive tolerance to heat, it is also necessary to test against individual and simultaneous stresses of heat and drought. Additionally, the tolerance to low-moisture or drought stress does not confirm tolerance to heat stress (Cairns *et al.*, 2013) and which is evident as the absence of correlation between drought tolerance and drought+heat tolerance (Meseka, Menkir, Bossey, & Mengesha, 2018).

Currently, very few studies are available on the effect of heat and heat+drought under managed stress conditions, vis-a-vis optimum conditions on the overall performance of the genotypes (Badu-Apraku & Fakorede, 2017; Cairns *et al.*, 2013; Meseka *et al.*, 2018; Nelimor *et al.*, 2019) studied the effect of individual and combined stresses *viz.*, drought stress, heat stress and drought+heat in the same location by eliminating confounding effects and also accounted for the compounding effects on grain yield while studying the individual and drought+heat stress separately. A very high correlation was observed between yield under heat stress and drought+heat stress indicating that the genotypes which withstand heat stress likely to withstand drought+heat stress as well. However, it has been observed that heat and heat+drought stress has a negative-additive effect on the overall grain yield of the genotypes but differs from the genetic background (Nelimor *et al.*, 2019). In contrast to the observations made by Cairns *et al.* (2013); Badu-apraku and Fakorede (Badu-Apraku & Fakorede, 2017) observed higher correlation for grain yield performance between drought stress (r=0.56) and heat+drought stress in the extra early white genotypes (r=0.26) as compared to extra early yellow genotypes, suggesting the role of genetic background on correlated response to heat and drought stress tolerance.

## Genetics of Heat Tolerance in Maize

Breeding of crops for any target trait(s) necessitates the knowledge on the variability and genetics of trait. Heat stress tolerance in maize is a complex phenomenon and influenced by numerous genes. The contribution of several genetic and environmental factors made the selection process complicated for physiologists and plant breeders. Few studies have been published on the inheritance pattern of heat tolerance in maize. Generation means analysis showed high broad- and narrow-sense heritability (>60%) for heat stress adaptive traits like leaf temperature and cell membrane thermo-stability and indicated the predominance of additive genetic variance in the inheritance of these adaptive traits (Naveed, Ahsan, Akram, Aslam, & Ahmed, 2016b). Further, genetic analysis of heat stress tolerance associated traits revealed the prevalence of non-additive gene action in the inheritance pattern of tassel blast and leaf firing (Jodage *et al.*, 2017). Similarly, significant specific combining ability (SCA) effects for membrane thermostability traits were recorded in spring maize (Kaur, Saxena, & Malhi, 2010). Additionally, in a heat stress environment, the epistatic interactions were reported for leaf firing, shelling percentage and membrane thermostability (Kaur & Saxena, 2012). Chen *et al.* (2012) reported the dominant inheritance of leaf firing and tassel blast traits owing to the absence of heat-induced leaf firing and tassel blast phenotypes in the hybrid plants made from a heat-tolerant parent and a heat-sensitive parent. Instead, tassel blasting and leaf firing phenotypes were expressed in the hybrids derived from heat-sensitive parents (B106 × NC350 and NC350 × B106). The functional genomics approach was employed to understand the gene action for heat tolerance in maize. Transcriptome analysis in maize hybrid Annong 591 and its parental lines, CB25 and CM1 in response to heat stress, showed a preponderance of non-additive gene action for more the 5400 gene expression (Zhao *et al.*, 2019).

## Conventional Breeding

Heritable variability for the target trait in the crops paved the foundation for the genetic improvement of crops. Mazie holds high genetic variability for heat stress adaptive traits (Alam *et al.*, 2017; Naveed *et al.*, 2016a, 2016b). However, owing to the continued natural and artificial selection processes, germplasm from warmer ecological zones are relatively more heat tolerant than of cool regions (Yamamoto *et al.*, 2011; Yousaf *et al.*, 2018). Conventional breeding efforts for the development of heat-tolerant maize inbred lines mainly employ two different strategies; firstly, screening of the existing maize germplasm and elite maize hybrids for heat tolerance and secondly planned and targeted breeding efforts for the development of heat-tolerant maize germplasm and hybrids.

Screening of tropical maize inbred at the reproductive stage under heat stress revealed nine heat-tolerant maize inbred lines *viz.*, INB14, INB27, INB28, INB32, INB39, INB49, INB67, INB74 and INB78 (Noor *et al.*, 2019). In Pakistan, evaluation of exotic and local maize hybrids under heat stress revealed locally bred heat-tolerant hybrids, FH-988, FH-922, FH-1046 and YH-1898 with maximum grain yield and quality traits (Yousaf *et al.*, 2018). Further, field screening of seventy-five diverse tropical maize inbred lines from the CIMMYT-Asia maize program under natural heat stress condition identified VL05728 and VL05799 genotypes showing better reproductive success (pollination and fertilisation) under heat stress as tolerant lines (Alam *et al.*, 2017). Exposure to heat stress in field and greenhouse condition in Texas, USA revealed several inbred lines (B76, Tx205, C273A, BR1, B105C, C32B, S1W, and C2A554-4) with heat tolerance during reproductive stages; while inbred lines B106 and NC350 showed severe leaf firing and tassel blasting indicating high susceptibility (J. Chen *et al.*, 2012). The studies on variability for cell membrane thermo-stability and leaf temperature among one hundred maize inbred lines at mean day/night temperatures of 33°C/19°C (field) and 40°C/23°C (plastic tunnel) with optimum irrigation resulted in two efficient heat-tolerant inbred lines, ZL-11271 and A545 for both the targeted traits (Naveed *et al.*, 2016b).

The line × testers analysis in tropical maize inbred lines for grain yield under heat stress revealed several potential hybrids *viz.*, L118 × L2, L143 × L1 (Akula *et al.*, 2016), VL1011 × VL1110175, ZL126643 × VL1010877, VL126643 × VL0556, VL108868 × VL1110175 (Gazala *et al.*, 2017), and VL107 × VL128 (Archana *et al.*, 2018). The 300 testcross progenies developed by crossing the broad-based CML539 with 300 inbred lines showing the variable level of tolerance to different stresses [soil acidity, pest resistance, drought, low N and striga] and yield potential were evaluated against drought and heat stresses. The study identified two high yielding maize inbreds La Posta Seq C7-F64-2-6-2-2 and DTPYC9-F4--1-2-1-2, under both drought and combined drought and heat stress (Cairns *et al.*, 2013). Studies on combining ability for heat tolerance in spring maize revealed inbred lines I137, I132 and SE 527 as good general combiners and CM124 × (FR 632H-100-7-3 × M017) showed better significant SCA effects for membrane thermostability across environments (Kaur *et al.*, 2010). Pollens are highly sensitive to high-temperature stress. Taking advantage of pollen sensitivity to heat stress, gamete selection was employed for improving the heat tolerance in maize germplasm. The gamete selection was practiced in 'Arizona Arid Environment Maize' synthetic population. The study showed that the exposure of germplasm to elevated temperature during gamete function enriching the frequencies of genes associated with heat stress tolerance (Petolino, Cowen, Thompson, & Mitchell, 1990). Gamete selection for heat stress tolerance improved the grain yield by 17% and reduced the root lodging by 43% compared to non-selected progenies (Petolino, Cowen, Thompson, & Mitchell, 1992).

The CIMMYT has developed several stress-tolerant maize hybrids under the global maize breeding programme. Several donor inbred lines were identified and validated for heat tolerance under various breeding CIMMYT programs. Further, series of heat-tolerant maize hybrids were licensed to public and private sector institutions in India, Nepal, Bangladesh and Pakistan under the umbrella of project "Heat Stress Tolerant Maize for Asia" (https://htma.cimmyt.org). The information on some of the heat-tolerant maize genotypes is tabulated in Table **1**.

**Table 1. Representative tabular compilation of heat-tolerant maize germplasm** *viz.*, **landraces, inbred lines and hybrids identified for different regions and countries.**

| S. No. | Name Tolerant Germplasm | Type of Germplasm | Stage of Stress | Region | Reference |
|---|---|---|---|---|---|
| 1. | GH-4859, TZm-1353 | Landraces | Flowering | Sub Saharan Africa | Nelimor *et al.,* 2019 |
| 2. | INB14, INB27, INB28, INB32, INB39, INB49, INB67, INB74, INB78 | Inbred lines | Flowering & Early grain Filling | Peninsular India | Noor *et al.,* 2019 |
| 3. | S058, S067, L043, L012 | Inbred lines | Seedling | Germany | Frey *et al.,* 2015 |
| 4. | B73 and B76 | Inbred lines | Flowering | Texas, USA | Chen *et al.,* 2012 |
| 5. | VL05728 and VL05799 | Inbred lines | Reproductive | Peninsular India | Alam *et al.,* 2017 |
| 6. | K166B, K166A | Inbred lines | Pollination & Grain filling | Khuzestan (Iran) | Khodarahmpour, Choukan, Bihamta, & Majidi Hervan, 2011 |
| 7. | CML304, CML 305, CML306, CML18, CML161, CML172, CML189, CML308, CML451 | Inbred lines | Flowering | Eastern India | Rani, R.B.P, Kumari, Singh, & Kumari, 2018 |
| 8. | ZL135005, CAL1730, ZL132088, CZL0522 | Inbred lines | Flowering & Grain Filling | Peninsular India | Jodage *et al.,* 2018 |
| 9 | I137, I132, CM140, CM124, SE527, CM125 | Inbred lines | Flowering & Grain Filling | North-Western region, India | Kaur *et al.,* 2010 |
| 10. | L6-Y, L24-Y and Sweety 015 | Inbred line | Seedling | South Africa | Tandzi *et al.,* 2019 |

*(Table 1) cont.....*

| S. No. | Name Tolerant Germplasm | Type of Germplasm | Stage of Stress | Region | Reference |
|---|---|---|---|---|---|
| 11. | RML91, RML140, RML76, RML40 | Inbred line | Flowering & Grain Filling | Nepal | Manoj, Ghimire, Ojha, & Shrestha, 2018 |
| 12. | FH988, FH922, FH1046 & YH1898 | Hybrid | Flowering & Grain Filling | Sahiwal (Pakistan) | Yousaf *et al.*, 2018 |
| 13. | YH-1898, KJ. Surabhi, FH-793 ND-6339 and NK - 64017 | Hybrids | Flowering | Pakisthan | Rahman *et al.*, 2013 |
| 14. | K18 × K166B | Hybrid | Pollination & Grain filling | Khuzestan (Iran) | Khodarahmpour *et al.*, 2011 |
| 15. | B76 × NC350, B76 × B106 | Hybrids | Flowering | Texas, USA | Chen *et al.*, 2012 |
| 16. | L118 × L12, L143 × L1 | Hybrids | Flowering & Grain Filling | Peninsular India | Akula *et al.*, 2016 |
| 17. | I137 × I155, I137 × SE527, | Hybrids | Flowering & Grain Filling | North-Western region, India | Kaur *et al.*, 2010 |
| 18. | La posta Sequia C7-F64-2-6-2-2 and DTpYC9-F4--1-2-1-2, | Test-crosses with CML539 | Reproductive | Mexico, Kenya, India, and Zimbabwe, Thailand | Cairns *et al.*, 2013 |
| 19. | M1227-17, M0826-3, and M1124-18 | Three-way cross hybrids | Flowering & Grain Filling | West and Central Africa | Meseka *et al.*, 2018 |

## Genomics Assisted Breeding for Heat Tolerance

Advances in genotyping platforms and informatics revolutionised the trait mapping in crop plants. The QTL mapping strategy was applied to decipher the genetic relationship among tolerance to various stresses in major crops, including maize (Collins, Tardieu, & Tuberosa, 2008). However, there are few reports in maize with respect to the dissection of marker-trait association of heat tolerance associated traits under heat stress. In maize, initial efforts to map the thermotolerance QTL for pollen-associated traits, cellular membrane stability, quantitative expression of HSP and radicle growth were made with RFLP markers (Frova, 1996; Frova, Caffulli, & Pallavera, 1998; Frova & Sari-Gorla, 1994). The studies detected five QTL for pollen germination and six QTL for pollen tube growth. Further, QTL were identified for the heat-sensitive phenological and grain traits *viz.*, time to anthesis and silking, ASI, leaf scorching, dry grain yield in six $F_{3:5}$ populations derived using four dents and four flint inbred lines. The study showed six heat-tolerance and 112 heat-responsive candidate genes co-located with the previously known heat-tolerant QTL. In addition to leaf firing, leaf blotching and tassel blasting are the relevant visible heat stress indices in maize

(Frey *et al.*, 2016). In a similar mapping population, six QTL explaining 7 and 9% of the phenotypic variance were identified for heat susceptibility index (HSI) associated traits *viz.*, leaf length, plant height, leaf scorching, leaf greenness and leaf growth rate (Inghelandt *et al.*, 2019). (McNellie *et al.*, 2018) detected the 12, 8 and 1 QTL for leaf firing, leaf blotching and tassel blasting, respectively, in the bi-parental mapping populations derived from B73 × NC350 and B73 × CML103. Further, the study also revealed the major QTL for heat-induced plant death on chromosome 3 with a phenotypic variation of 26.2%. A recent study by Gao *et al.* (2009) revealed the QTL for thermotolerance of seed set using bi-parental and association mapping approach. From the linkage mapping, four QTL were identified, and 42 SNPs distributed among 17 genes were detected through GWAS. Among 17 genes, four candidate genes were co-localized with the QTL identified in linkage mapping. The GWAS in 300 tropical and subtropical maize inbred lines revealed 6, 36 and 55 candidate genes for anthesis date, ASI and grain yield, respectively (Yuan *et al.*, 2019). The details of the QTL mapping studies for heat tolerance in maize are summarised in Table **2**.

**Table 2. Details of QTL mapping studies conducted in maize for heat tolerance associated traits.**

| S. No. | Population/Panel | Target Traits | Genotyping | No. of QTL | PVE [%] | Country of Phenotyping | Phenotyping Environment | Reference |
|---|---|---|---|---|---|---|---|---|
| 1. | RILs<br>• BT-1 × N6 (N=237),<br>• GWAS (N=261) | Thermotolerance of seed-set | GBS | RILs: 4<br>GWAS: 42 | 10.20-11.30<br>6.50–9.70 | China | Field | Gao *et al.*, 2019 |
| 2*. | GWAS (N=300) | Anthesis date, ASI, and grain yield | GBS | GWAS: 97 | - | Mexico, Kenya, Thailand, Zimbabwe and India | Field | Yuan *et al.*, 2019 |
| 3. | RILs<br>• (B73 × NC350; N= 185)<br>• (B73 × CML103; N= 195) | Leaf firing, leaf blotching, tassel blasting, reduction in spikelet size, plant death. | SNP (N=1144) | 22 | 6.54-26.21 | USA | Field | McNellie *et al.*, 2018 |

*(Table 2) cont.....*

| S. No. | Population/Panel | Target Traits | Genotyping | No. of QTL | PVE [%] | Country of Phenotyping | Phenotyping Environment | Reference |
|---|---|---|---|---|---|---|---|---|
| 4. | Six $F_{3:5}$ populations • S067 ×P040 • S067 × L012 • L012×L017 • L043 ×L023 • S070 ×L023 S058 × S070 (N=607) | Leaf growth rate (LR), leaf greenness (SD), leaf scorching of young leaves (SC), leaf senescence of old leaves (SN), leaf length (LL), Plant height (PH), number of leaves (NL), shoot dry weight (DW), and shoot water content (WC). | SNP (N=170) | 7 | 7.00-13.00 | Germany | Controlled | Inghelandt et al., 2019 |
| 5. | Six $F_{3:5}$ populations • S067 ×P040 (N=107) • S067 × L012 (N=107) • L012×L017 (N=106) • L043 ×L023 (N=106) • S070 ×L023 (N=75) • S058 × S070 (N=107) | ASI, leaf scorching, dry grain yield, time to female and male flowering, and grain moisture | SNP (N = 161) | 11 | 7.00-13.00 | Germany Hungary Italy | Field | Frey et al., 2016 |
| 6*. | RILs • T232 × CM37 (N=48) • B73 × H99 (N=142) | Cell membrane stability (CMS), quantitative expression of heat shock proteins (qHSP), leaf water retention, root growth, pollen germination (PG), tube growth (TG) | RFLP (N=200) SSR (N=70) | 45 | 35.00-88.00 | Italy | Controlled | Frova et al., 1998 |
| 7. | RILs • T232 × CM37 (N= 45) | CMS, qHSP, Radicle growth, PG and TG | RFLP (N=200) | 45 | 35.00-88.00 | Italy | Controlled | Frova, 1996 |
| 8. | RILs (T232 × CM37, N= 45) | PG, TG, thermotolerance of germinability and thermotolerance of tube growth | RFLP (N=184) | 55[#] | 8.00-42.00 | Italy | Controlled | Frova & Sari-Gorla, 1994 |

*Study includes for both water and heat stress; #Excluding repetitive loci for multiple traits.

The use of identified QTL in MAS is limited to major QTL and omitted minor QTL in the selection process resulting in loss of genetic gain (Dekkers, 2004). To avoid these limitations, genomic selection (GS) has been proposed as a method to understand the effects of all the alleles across the genome to improve polygenic traits (Meuwissen & Goddard, 2010; Shikha *et al.*, 2017). Very few studies have undertaken to predict the genomic estimated breeding values during heat stress. Trachsel *et al.*, (2019) employed genomic selection models *viz.*, partial least squares regression (PLSR), random forest (RF), ridge regression (RR) and Bayesian ridge regression (BayesB) to predict the physiological genomic estimated breeding values (PGEBV) for grain yield in combined drought and heat stress environments. Highest prediction accuracy within and across location predictions (rGP) were observed for BayesB followed by RR, RF and PLSR. The genomic selection for grain yield, an thesis date and ASI revealed the genomic prediction accuracies (rMG) of 0.69, 0.61 and 0.28, respectively for marker trait-associated SNPs (Yuan *et al.*, 2019). The prediction accuracies of genome-wide prediction models for heat tolerance were found high within the population rather than among or mixed genetic groups (Inghelandt *et al.*, 2019). Additionally, Li *et al.*, (2018) recently showed that the upstream region of *bZIP60* is essential in conditioning the response to heat stress in maize.

## Genetic Engineering

Advances in genetic transformation and gene expression techniques in maize allowed rapid development of transgenics cultivars through genetic engineering and functional characterisation of genes for heat tolerance. The genetic engineering approaches were employed to validate the maize genes for thermotolerance in maize and model plants and to develop the maize transgenics lines.

Several maize genes were validated in model and crop systems for the ability to withstand heat stress. Expression of maize HSP genes, *ZmHSP16*.9 and *ZmHsf06*, enhanced the heat and drought-stress tolerance in tobacco and Arabidopsis as compared to wild plants (Li *et al.*, 2015; Sun *et al.*, 2012). *Arabidopsis* transgenic lines transformed with *ZmHsf04* showed enhanced thermotolerance by upregulating the expression of native HSP, and stress-related genes (*AtHsp25*.3-P, *AtHsp18*.2-CI, *AtHsp70B*, *AtAPX2* and *AtGolS1*) compared to the wild type (Jiang, Zheng, Chen, Liang, & Wu, 2018). The unregulated expression of *ZmHsf12* was reported in many maize organs with response heat shock treatment. Further, *ZmHsf12* expression in yeast cells and *Arabidopsis* background significantly enhanced both basal and acquired thermotolerance (Li *et al.*, 2019a). The new heat shock transcription factor *Hsf05* cloned from maize, and its overexpression in *Arabidopsis thaliana* improved the tolerance level to heat stress (Li *et al.*, 2019b). Similarly, overexpression of *ZmHSFA2* in *Arabidopsis* increased the raffinose content in leaves through enhanced expression of *AtRS5* [Raffinose synthase] and resulted in heat stress tolerance (Gu *et al.*, 2019).

Maize plants transformed with rice MYB transcription factor gene *OsMYB55* improved the tolerance for heat and drought stresses *via* modulating the expression of stress-associated genes (Casaretto *et al.*, 2016). Overexpression of maize WRKY transcription factor *ZmWRKY106* improved the tolerance to drought and heat in transgenic *Arabidopsis* by regulating expression patterns of genes associated with ABA-signalling pathway and superoxide dismutase (SOD), peroxide dismutase (POD), and catalase (CAT) (Wang *et al.*, 2018). The bZIP factors serve as positive regulators of abiotic stress responses in maize. Overexpression of native bZIP factor *ZmbZIP4* in maize provided the tolerance for abiotic stresses *viz.*, drought, cold and heat by improving the root system architecture and abscisic acid synthesis (Ma *et al.*, 2018). Recently, modification of heat resistant cytoplasmic heat-stable 6-phosphogluconate dehydrogenase (6PGDH) to target the plastids resulted in heat tolerance in transgenic maize (Patent No. WO 2019/090265 Al, 2019).

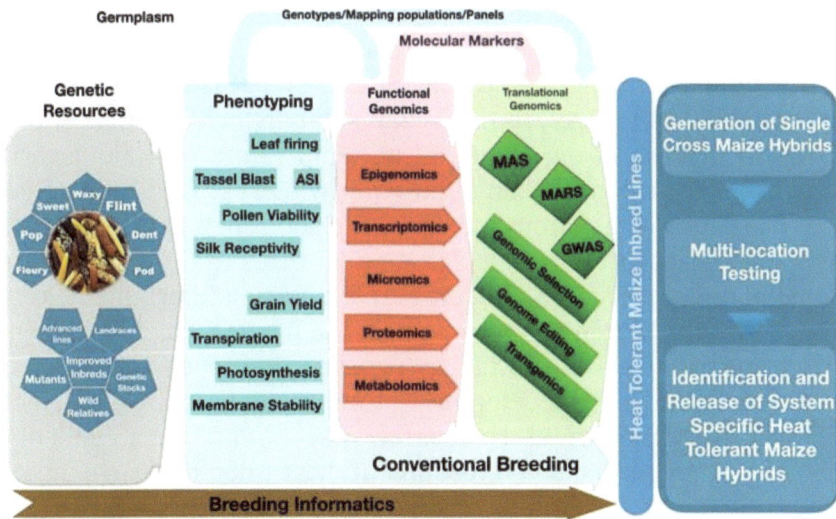

**Fig. (3).** Breeding pipeline for developing system specific heat stress tolerant maize hybrids using conventional and next generation breeding tools.

There are pros and cons in relation to cost involved, anticipated benefits, the quantum of risk involved, probability or percent success for each of the breeding techniques depending upon the resource requirement, technicity, trait complexity and nature of germplasm. For example, the development of heat-tolerant genotypes by employing conventional breeding methods would be a generally slow process, which requires less resources and infrastructure. On the contrary, employing genomic selection or transgenic technology would be faster but requires state-of-art facilities. Further, the use of any specific breeding approach relies on target trait(s), available variability, phenotyping tools, resources and skilled human sources. Breeding efficiency for heat tolerance starts with robust phenotyping to the adaptation of most feasible breeding methodologies. The generalized breeding pipeline for heat tolerance in maize using conventional and modern breeding tools is showed pictorially in Fig. (**3**).

## PROSPECTS

The world population is anticipated to reach 9 billion by 2050 and which forces farmers to enhance the global maize production from 1147 million tonnes (2018) to 1690 million tonnes (2050). Similarly, there is a high pressure to enhance the Indian maize production from the existing 27.82 million tonnes to 121 million tonnes by 2050 to meet the requirement of growing population (Amarasinghe, Shah, & Singh, 2007; Raju *et al.*, 2018). However, heat stress is going to be a significant challenge for achieving the target maize production in the era of global warming and climate change. The average mean global temperature expected to rise by 2°C by 2050. Similarly, in the Indian sub-continent, Murari *et al.* (2001) anticipating an increase in the annual mean temperature of 3.5°C and 5.5°C by 2080, during *kharif* [monsoon] than in the *rabi* [winter] season, respectively. The majority of maize growing regions of tropics and subtropics are the most sensitive areas for heat stress occurrence. The substantial efforts have been undertaken by international and national maize research fraternity to address the heat stress to sustain the maize production and to safeguard the maize farmers and consumers. However, growing demand for maize consumption necessitates continuous scientific efforts towards the development of improved heat stress-tolerant maize cultivars with higher grain yield. Hence, few potential options to understand the heat tolerance mechanisms and breeding of heat-tolerant maize hybrids are discussed briefly under the following headings.

## High Throughput and Reliable Phenotyping

The success of any breeding program relies on the stringent phenotyping for target traits. Precise phenotyping ends up with an accurate selection. Being a quantitative trait, heat stress tolerance is governed by multiple adaptive mechanisms. Therefore, grouping tolerant and susceptible inbred lines based on the presence or absence of adaptive trait(s) will greatly help in choosing of genotypes for trait introgression. Many phenotyping methods at the seedling stage will not provide direct information on heat stress on grain yield parameters. Similarly, the screening of maize inbred lines at the reproductive stage requires comparatively more resources and manpower. Therefore, there is a need to device the phenotyping approaches, which are more informative, reliable and high throughput in nature.

## Deciphering the Interactions Between Heat and Other Abiotic Stresses on Maize

The severity of climate change is expected to produce more pronounced effects on maize grain yield when there are simultaneous occurrences of multiple abiotic stresses. The occurrence of heat stress mostly coincides combines with drought. Therefore, studying the effects of multiple-stresses on phenology and grain yield in different maize germplasms and understanding the genetic and molecular basis of multi-stress interactions facilitate the breeders to come out with multi stress-tolerant maize hybrids.

## Dissecting Detailed Molecular Mechanisms and Signalling Cascades of Heat Stress Tolerance in Maize

The molecular basis of heat tolerance was majorly dissected in model plants like *Arabidopsis*. The efforts need to be channelized to understand the molecular mechanisms of each of the functional adaptive traits of heat tolerance in maize. Further, the development of genomic atlas at various expression strata and combining the results to understand systems biology of heat tolerance is another important area to be looked into. Additionally, owing to the geographical domestication, there is a variable level of tolerance in different groups of maize germplasm. Therefore, understanding of the genetic and molecular bases of the variable response of different maize groups towards heat stress helps in the greater utility of maize germplasm in heat stress breeding programs.

## Use of Next-generation Breeding Tools

Advances in genomics and molecular biology come up with various novel breeding tools accompanied by enhanced accuracy and genetic gain per unit time. The rapid advancement of generations through speed breeding and doubled haploid technology are promising approaches to develop heat-tolerant maize hybrids within a limited time scale (Chaikam, Molenaar, Melchinger, &

Boddupalli, 2019; Ghosh *et al.*, 2018). The genome editing tools allowing the precise editing of target locations in the genomic regions are the most promising in maize breeding programmes. Presently, genome editing was found successful for complex traits like drought tolerance in maize. Shi *et al.* (2017) developed drought-tolerant cultivars through the editing of *ARGOS8*. Therefore, there is ample scope for the employment of genome editing strategies for targeting heat stress associated with key candidate genes and regulatory elements to develop heat-tolerant maize hybrids.

## Policy Support

Policy support to undertake and promote the climate-resilient breeding programs is utmost needed, especially in the developing and climate change prone countries. Establishment of high throughput facilities to undertake the large-scale screening for heat and other abiotic stresses and would help in the phenotyping of large set of maize germplasm and rapid advancement of breeding cycles. Further, regulating the release of maize cultivars with a minimum level of heat tolerance could ensure the grain yield in heat stress affected maize growing areas and during the occurrence of unwarranted heat waves.

## CONSENT FOR PUBLICATION

Not applicable.

## CONFLICT OF INTEREST

The authors confirm that this chapter content has no conflict of interest.

## ACKNOWLEDGEMENTS

We thank the ICAR project on Computational Biology and Agricultural Bioinformatics (Agril.Edn.14 (44)/2014-A&P) for funding the research work on maize abiotic stress tolerance. The funders had no role in study design, data collection and analysis, decision to publish, or preparation of the manuscript.

## REFERENCES

Ainsworth, E.A., Ort, D.R. (2010). How do we improve crop production in a warming world? *Plant Physiol.*, *154*(2), 526-530.
[http://dx.doi.org/10.1104/pp.110.161349] [PMID: 20921178]

Agrawal, P.K., Mallikarjuna, M.G., Gupta, H.S. (2018). Genetics and applied genomics of quality protein maize for food and nutritional security. *Biotechnologies of Crop Improvement.* (Vol. 3, pp. 151-178). Cham: Springer International Publishing.http://link.springer.com/10.1007/978-3-319-94746-4_7
[http://dx.doi.org/10.1007/978-3-319-94746-4_7]

Akula, D., Patil, A., Zaidi, P.H., Kuchanur, P.H., Vinayan, M.T., Seetharam, K. (2016). Line x testers analysis of tropical maize inbred lines under heat stress for grain yield and secondary traits. *Maydica, 61*(1), 3-6.

Alam, M.A., Seetharam, K., Zaidi, P.H., Dinesh, A., Vinayan, M.T., Nath, U.K. ( 2017). Dissecting heat stress tolerance in tropical maize (*Zea mays* L.). *F. Crop Res., 204*, 9-110.
[http://dx.doi.org/http://dx.doi.org/10.1016/j.fcr.2017.01.006]

Almeselmani, M., Deshmukh, P.S., Sairam, R.K., Kushwaha, S.R., Singh, T.P. (2006). Protective role of antioxidant enzymes under high temperature stress. *Plant Sci., 171*(3), 382-388.
[http://dx.doi.org/10.1016/j.plantsci.2006.04.009] [PMID: 22980208]

Amarasinghe, U., Shah, T., Singh, O. (2007). *Changing Consumption Patterns: Implications on Food and Water Demand in India.* Colombo.

Angadi, S.V., Cutforth, H.W., Miller, P.R., McConkey, B.G., Entz, M.H., Brandt, S.A. (2000). Response of three brassica species to high temperature stress during reproductive growth. *Can. J. Plant Sci., 80*(4), 693-701.
[http://dx.doi.org/10.4141/P99-152]

Archana, K.A., Kuchanur, P., Zaidi, P.H., Mandal, S., Arunkumar, B., Patil, A. (2018). Combining ability and heterosis in tropical maize (*Zea mays* L.) under heat stress. *J. Pharmacogn. Phytochem., 7*(6), 1026-1031.

Arora, K., Panda, K.K., Mittal, S., Mallikarjuna, M.G., Rao, A.R., Dash, P.K., Thirunavukkarasu, N. (2017). RNAseq revealed the important gene pathways controlling adaptive mechanisms under waterlogged stress in maize. *Sci. Rep., 7*(1), 10950.
[http://dx.doi.org/10.1038/s41598-017-10561-1] [PMID: 28887464]

Asada, K. (2006). Production and scavenging of reactive oxygen species in chloroplasts and their functions. *Plant Physiol., 141*(2), 391-396.
[http://dx.doi.org/10.1104/pp.106.082040] [PMID: 16760493]

Ashraf, M., Hafeez, M. (2004). Thermotolerance of pearl millet and maize at early growth stages: Growth and nutrient relations. *Biol. Plant., 48*(1), 81-86.
[http://dx.doi.org/10.1023/B:BIOP.0000024279.44013.61]

Asthir, B. (2015). Protective mechanisms of heat tolerance in crop plants. *J. Plant Interact., 10*(1), 202-210.
[http://dx.doi.org/10.1080/17429145.2015.1067726]

Badu-Apraku, B., Fakorede, M.A.B. (2017). Improvement of early and extra-early maize for combined tolerance to drought and heat stress in Sub-Saharan Africa.*Advances in Genetic Enhancement of Early and Extra-Early Maize for Sub-Saharan Africa.* (pp. 311-358). Cham: Springer International Publishing.http://link.springer.com/10.1007/978-3-319-64852-1_12
[http://dx.doi.org/10.1007/978-3-319-64852-1_12]

Balla, K., Rakszegi, M., Li, Z., Békés, F., Bencze, S., Veisz, O. (2011). Quality of winter wheat in relation to heat and drought shock after anthesis. *Czech J. Food Sci., 29*(2), 117-128.
[http://dx.doi.org/10.17221/227/2010-CJFS]

Bassu, S., Brisson, N., Durand, J.L., Boote, K., Lizaso, J., Jones, J.W., Rosenzweig, C., Ruane, A.C., Adam, M., Baron, C., Basso, B., Biernath, C., Boogaard, H., Conijn, S., Corbeels, M., Deryng, D., De Sanctis, G., Gayler, S., Grassini, P., Hatfield, J., Hoek, S., Izaurralde, C., Jongschaap, R., Kemanian, A.R., Kersebaum, K.C., Kim, S.H., Kumar, N.S., Makowski, D., Müller, C., Nendel, C., Priesack, E., Pravia, M.V., Sau, F., Shcherbak, I., Tao, F., Teixeira, E., Timlin, D., Waha, K. (2014). How do various maize crop models vary in their responses to climate change factors? *Glob. Change Biol., 20*(7), 2301-2320.
[http://dx.doi.org/10.1111/gcb.12520] [PMID: 24395589]

Berry, J., Bjorkman, O. (1980). Photosynthetic response and adaptation to temperature in higher plants. *Annu. Rev. Plant Physiol., 31*(1), 491-543.
[http://dx.doi.org/10.1146/annurev.pp.31.060180.002423]

Begcy, K., Nosenko, T., Zhou, L-Z., Fragner, L., Weckwerth, W., Dresselhaus, T. (2019). Male sterility in

maize after transient heat stress during the tetrad stage of pollen development. *Plant Physiol., 181*(2), 683-700.
[http://dx.doi.org/10.1104/pp.19.00707] [PMID: 31378720]

Ben-Asher, J., Garcia, Y., Garcia, A., Hoogenboom, G. (2008). Effect of high temperature on photosynthesis and transpiration of sweet corn (*Zea mays* L. var. rugosa). *Photosynthetica, 46*(4), 595-603.
[http://dx.doi.org/10.1007/s11099-008-0100-2]

Bhadula, S.K., Elthon, T.E., Habben, J.E., Helentjaris, T.G., Jiao, S., Ristic, Z. (2001). Heat-stress induced synthesis of chloroplast protein synthesis elongation factor (EF-Tu) in a heat-tolerant maize line. *Planta, 212*(3), 359-366.
[http://dx.doi.org/10.1007/s004250000416] [PMID: 11289600]

Bokszczanin, K.L., Fragkostefanakis, S. Solanaceae pollen thermotolerance initial training network (SPOT-ITN) consortium. (2013). Perspectives on deciphering mechanisms underlying plant heat stress response and thermotolerance. *Front. Plant Sci., 4*(AUG), 315.
[http://dx.doi.org/10.3389/fpls.2013.00315] [PMID: 23986766]

Caers, M., Rudelsheim, P., Van Onckelen, H., Horemans, S. (1985). Effect of heat stress on photosynthetic activity and chloroplast ultrastructure in correlation with endogenous cytokinin concentration in maize seedlings. *Plant Cell Physiol., 26*(1), 47-52.

Cairns, J.E., Crossa, J., Zaidi, P.H., Grudloyma, P., Sanchez, C., Luis Araus, J. (2013). Identification of drought, heat, and combined drought and heat tolerant donors in maize. *Crop Sci., 53*(4), 1335-1346.
[http://dx.doi.org/10.2135/cropsci2012.09.0545]

Cairns, J.E., Sonder, K., Zaidi, P.H., Verhulst, N., Mahuku, G., Babu, R. ( 2012). Maize production in a changing climate. impacts, adaptation, and mitigation strategies. *Adv. Agron., 114*, 1- 58.

Cairns, J.E., Sanchez, C., Vargas, M. (2012). Dissecting maize productivity: Ideotypes associated with grain yield under drought stress and well-watered conditions. *J. Integr. Plant. Biol., 54*(12), 20.
[http://dx.doi.org/10.1111/j.1744-7909.2012.01156.x]

Casaretto, J.A., El-Kereamy, A., Zeng, B., Stiegelmeyer, S.M., Chen, X., Bi, Y-M., Rothstein, S.J. (2016). Expression of *OsMYB55* in maize activates stress-responsive genes and enhances heat and drought tolerance. *BMC Genomics, 17*, 312.
[http://dx.doi.org/10.1186/s12864-016-2659-5] [PMID: 27129581]

Collins, N.C., Tardieu, F., Tuberosa, R. (2008). Quantitative trait loci and crop performance under abiotic stress: where do we stand? *Plant Physiol., 147*(2), 469-486.
[http://dx.doi.org/10.1104/pp.108.118117] [PMID: 18524878]

Chaikam, V., Molenaar, W., Melchinger, A.E., Boddupalli, P.M. (2019). Doubled haploid technology for line development in maize: Technical advances and prospects. *Theor. Appl. Genet., 132*(12), 3227-3243.
[http://dx.doi.org/10.1007/s00122-019-03433-x] [PMID: 31555890]

Cooper, P., Rao, K., Singh, P., Dimes, J., Traore, P., Rao, K. (2009). Farming with current and future climate risk: Advancing a "Hypothesis of Hope" for rainfed agriculture in the semi-arid tropics. *J. SAT Agric. Res., 7*(December), 1-19.

Cicchino, M., Rattalino Edreira, J.I., Uribelarrea, M., Otegui, M.E. (2010). Heat stress in field-grown maize: Response of physiological determinants of grain yield. *Crop Sci., 50*(4), 1438-1448.
[http://dx.doi.org/10.2135/cropsci2009.10.0574]

Crafts-Brandner, S.J., Salvucci, M.E. (2002). Sensitivity of photosynthesis in a C4 plant, maize, to heat stress. *Plant Physiol., 129*(4), 1773-1780.http://www.plantphysiol.org/lookup/doi/10.1104/pp.002170
[http://dx.doi.org/10.1104/pp.002170] [PMID: 12177490]

Chen, Y., Burris, J.S. (1991). Desiccation tolerance in maturing maize seed: Membrane phospholipid composition and thermal properties. *Crop Sci., 31*(3), 766.
[http://dx.doi.org/10.2135/cropsci1991.0011183X003100030046x]

Cheikh, N., Jones, R.J. (1994). Disruption of maize kernel growth and development by heat stress (role of

cytokinin/abscisic acid balance). *Plant Physiol., 106*(1), 45-51.
[http://dx.doi.org/10.1104/pp.106.1.45] [PMID: 12232301]

Chen, J., Xu, W., Velten, J., Xin, Z., Stout, J. (2012). Characterization of maize inbred lines for drought and heat tolerance. *J. Soil Water Conserv., 67*(5), 354-364.
[http://dx.doi.org/10.2489/jswc.67.5.354]

Davletova, S., Rizhsky, L., Liang, H., Shengqiang, Z., Oliver, D.J., Coutu, J., Shulaev, V., Schlauch, K., Mittler, R. (2005). Cytosolic ascorbate peroxidase 1 is a central component of the reactive oxygen gene network of Arabidopsis. *Plant Cell, 17*(1), 268-281.
[http://dx.doi.org/10.1105/tpc.104.026971] [PMID: 15608336]

Dekkers, J.C.M. (2004). Commercial application of marker- and gene-assisted selection in livestock: strategies and lessons. *J. Anim. Sci., 82 E-Suppl*, E313-E328.
[PMID: 15471812]

Dinler, B.S., Demir, E., Kompe, Y.O. (2014). Regulation of auxin, abscisic acid and salicylic acid levels by ascorbate application under heat stress in sensitive and tolerant maize leaves. *Acta Biol. Hung., 65*(4), 469-480.
[http://dx.doi.org/10.1556/ABiol.65.2014.4.10] [PMID: 25475985]

Dresselhaus, T., Franklin-Tong, N. (2013). Male-female crosstalk during pollen germination, tube growth and guidance, and double fertilization. *Mol. Plant, 6*(4), 1018-1036.
[http://dx.doi.org/10.1093/mp/sst061] [PMID: 23571489]

Dupuis, I., Dumas, C. (1990). Influence of temperature stress on *in vitro* fertilization and heat shock protein synthesis in maize (*Zea mays* L.) reproductive tissues. *Plant Physiol., 94*(2), 665-670.
[http://dx.doi.org/10.1104/pp.94.2.665] [PMID: 16667763]

Eichten, S.R., Springer, N.M. (2015). Minimal evidence for consistent changes in maize DNA methylation patterns following environmental stress. *Front. Plant Sci., 6*(MAY), 308.
[http://dx.doi.org/10.3389/fpls.2015.00308] [PMID: 25999972]

Fahad, S., Bajwa, A.A., Nazir, U., Anjum, S.A., Farooq, A., Zohaib, A., Sadia, S., Nasim, W., Adkins, S., Saud, S., Ihsan, M.Z., Alharby, H., Wu, C., Wang, D., Huang, J. (2017). Crop production under drought and heat stress: Plant responses and management options. *Front. Plant Sci., 8*(June), 1147.
[http://dx.doi.org/10.3389/fpls.2017.01147] [PMID: 28706531]

Frey, F.P., Urbany, C., Hüttel, B., Reinhardt, R., Stich, B. (2015). Genome-wide expression profiling and phenotypic evaluation of European maize inbreds at seedling stage in response to heat stress. *BMC Genomics, 16*(1), 123.
[http://dx.doi.org/10.1186/s12864-015-1282-1] [PMID: 25766122]

Frova, C., Sari-Gorla, M. (1994). Quantitative trait loci (QTLs) for pollen thermotolerance detected in maize. *Mol. Gen. Genet., 245*(4), 424-430.http://link.springer.com/10.1007/BF00302254
[http://dx.doi.org/10.1007/BF00302254] [PMID: 7808391]

Frova, C. (1996). Genetic dissection of thermotolerance in maize. In: Grillo, S., Leone, A., (Eds.), *Physical Stresses in Plants.* (pp. 31-38). Berlin, Heidelberg: Springer Berlin Heidelberg. http://link.springer.com/10.1007/978-3-642-61175-9_3
[http://dx.doi.org/10.1007/978-3-642-61175-9_3]

Frova, C., Caffulli, A., Pallavera, E. (1998). Mapping quantitative trait loci for tolerance to abiotic stresses in maize. *J. Exp. Zool., 282*(1–2), 164-170.
[http://dx.doi.org/10.1002/(SICI)1097-010X(199809/10)282:1/2<164::AID-JEZ18>3.0.CO;2-U]

Frey, F.P., Presterl, T., Lecoq, P., Orlik, A., Stich, B. (2016). First steps to understand heat tolerance of temperate maize at adult stage: identification of QTL across multiple environments with connected segregating populations. *Theor. Appl. Genet., 129*(5), 945-961.
[http://dx.doi.org/10.1007/s00122-016-2674-6] [PMID: 26886101]

Gao, J., Wang, S., Zhou, Z., Wang, S., Dong, C., Mu, C., Song, Y., Ma, P., Li, C., Wang, Z., He, K., Han, C.,

Chen, J., Yu, H., Wu, J. (2019). Linkage mapping and genome-wide association reveal candidate genes conferring thermotolerance of seed-set in maize. *J. Exp. Bot., 70*(18), 4849-4864.
[http://dx.doi.org/10.1093/jxb/erz171] [PMID: 30972421]

Galic, V., Franic, M., Jambrovic, A., Ledencan, T., Brkic, A., Zdunic, Z., Simic, D. (2019). Genetic correlations between photosynthetic and yield performance in maize are different under two heat scenarios during flowering. *Front. Plant Sci., 10*(April), 566.
[http://dx.doi.org/10.3389/fpls.2019.00566] [PMID: 31114604]

Garcia y Garcia, A., Abritta, M.A., Soler, C.M.T., Green, A. (2014). Water and heat stress: The effect on the growth and yield of maize and the impacts on irrigation water. *WIT Trans Ecol Environ., 185*, 77-87.
[http://dx.doi.org/10.2495/SI140081]

Gazala, P, Kuchanur, PH, Zaidi, PH, Arunkumar, B, Patil, A, Seetharam, K ( 2017). Combining ability and heterosis for heat stress tolerance in maize (*Zea mays* L.). *J. Farm. Sci., 30*( 3), 33-326.

Ghosh, S., Watson, A., Gonzalez-Navarro, O.E., Ramirez-Gonzalez, R.H., Yanes, L., Mendoza-Suárez, M., Simmonds, J., Wells, R., Rayner, T., Green, P., Hafeez, A., Hayta, S., Melton, R.E., Steed, A., Sarkar, A., Carter, J., Perkins, L., Lord, J., Tester, M., Osbourn, A., Moscou, M.J., Nicholson, P., Harwood, W., Martin, C., Domoney, C., Uauy, C., Hazard, B., Wulff, B.B.H., Hickey, L.T. (2018). Speed breeding in growth chambers and glasshouses for crop breeding and model plant research. *Nat. Protoc., 13*(12), 2944-2963.
[http://dx.doi.org/10.1038/s41596-018-0072-z] [PMID: 30446746]

Gong, M., Chen, B.O., Li, Z.G., Guo, L.H. (2001). Heat-shock-induced cross adaptation to heat, chilling, drought and salt stress in maize seedlings and involvement of $H_2O_2$. *J. Plant Physiol., 158*(9), 1125-1130.
[http://dx.doi.org/10.1078/0176-1617-00327]

Goraya, G.K., Kaur, B., Asthir, B., Bala, S., Kaur, G., Farooq, M. (2017). Rapid injuries of high temperature in plants. *J. Plant Biol., 60*(4), 298-305.
[http://dx.doi.org/10.1007/s12374-016-0365-0]

Gu, L., Jiang, T., Zhang, C., Li, X., Wang, C., Zhang, Y., Li, T., Dirk, L.M.A., Downie, A.B., Zhao, T. (2019). Maize HSFA2 and HSBP2 antagonistically modulate raffinose biosynthesis and heat tolerance in Arabidopsis. *Plant J., 100*(1), 128-142.
[http://dx.doi.org/10.1111/tpj.14434] [PMID: 31180156]

Guo, M., Liu, J-H., Ma, X., Luo, D-X., Gong, Z-H., Lu, M-H. (2016). The plant heat stress transcription factors (HSFs): Structure, regulation, and function in response to abiotic stresses. *Front. Plant Sci., 7*(February), 114.
[http://dx.doi.org/10.3389/fpls.2016.00114] [PMID: 26904076]

Hasanuzzaman, M., Nahar, K., Alam, M.M., Roychowdhury, R., Fujita, M. (2013). Physiological, biochemical, and molecular mechanisms of heat stress tolerance in plants. *Int. J. Mol. Sci., 14*(5), 9643-9684.
[http://dx.doi.org/10.3390/ijms14059643] [PMID: 23644891]

Hussain, H.A., Men, S., Hussain, S., Chen, Y., Ali, S., Zhang, S., Zhang, K., Li, Y., Xu, Q., Liao, C., Wang, L. (2019). Interactive effects of drought and heat stresses on morpho-physiological attributes, yield, nutrient uptake and oxidative status in maize hybrids. *Sci. Rep., 9*(1), 3890.
[http://dx.doi.org/10.1038/s41598-019-40362-7] [PMID: 30846745]

Hu, X., Wu, L., Zhao, F., Zhang, D., Li, N., Zhu, G., Li, C., Wang, W. (2015). Phosphoproteomic analysis of the response of maize leaves to drought, heat and their combination stress. *Front. Plant Sci., 6*, 298.http://www.frontiersin.org/Plant_Proteomics/10.3389/fpls.2015.00298/abstract
[http://dx.doi.org/10.3389/fpls.2015.00298] [PMID: 25999967]

Hu, X., Li, Y., Li, C., Yang, H., Wang, W., Lu, M. (2010). Characterization of small heat shock proteins associated with maize tolerance to combined drought and heat stress. *J. Plant Growth Regul., 29*(4), 455-464.
[http://dx.doi.org/10.1007/s00344-010-9157-9]

Hu, X., Yang, Y., Gong, F., Zhang, D., Zhang, L., Wu, L., Li, C., Wang, W. (2015). Protein sHSP26 improves chloroplast performance under heat stress by interacting with specific chloroplast proteins in maize

(*Zea mays*). *J. Proteomics, 115*, 81-92.
[http://dx.doi.org/10.1016/j.jprot.2014.12.009] [PMID: 25540934]

Jiang, Y., Zheng, Q., Chen, L., Liang, Y., Wu, J. (2018). Ectopic overexpression of maize heat shock transcription factor gene ZmHsf04 confers increased thermo and salt-stress tolerance in transgenic Arabidopsis. *Acta Physiol. Plant., 40*(1), 1-12.
[http://dx.doi.org/10.1007/s11738-017-2587-2]

Jodage, K., Kuchanur, P.H., Zaidi, P.H., Patil, A., Seetharam, K., Vinayan, M.T. (2018). Genetic analysis of heat adaptive traits in tropical maize (*Zea mays* L.). *Int. J. Curr. Microbiol. Appl. Sci., 7*(1), 3237-3246.
[http://dx.doi.org/10.20546/ijcmas.2018.701.387]

Jodage, K., Kuchanur, P.H., Zaidi, P.H., Patil, A., Seetharam, K., Vinayan, M.T. (2017). Genetic analysis of heat stress tolerance and association of traits in tropical maize (*Zea mays* L.). *Environ. Ecol., 35*(3C), 2354-2360.

Kaur, R., Saxena, V.K., Malhi, N.S. (2010). Combining ability for heat tolerance traits in spring maize (*Zea mays* L.). *Maydica, 55*, 195-199.

Kaur, R., Saxena, V.K. (2012). Genetics of heat tolerance traits in spring maize (*Zea mays* L.). *Agric. Sci. Dig., 32*(3), 181-186.

Kandel, M., Ghimire, S.K., Shrestha, J. (2018). Mechanisms of heat stress tolerance in maize. *Azarian J. Agri., 5*(1), 20-27. https://www.cabdirect.org/cabdirect/abstract/20183364741

Khanna, P., Kaur, K., Gupta, A.K. (2017). Root biomass partitioning, differential antioxidant system and thiourea spray are responsible for heat tolerance in spring maize. *Proc. Natl. Acad. Sci., India, Sect. B Biol. Sci., 87*(2), 351-359.
[http://dx.doi.org/10.1007/s40011-015-0575-0]

Khodarahmpour, Z., Choukan, R., Bihamta, M.R., Majidi Hervan, E. (2011). Determination of the best heat stress tolerance indices in maize (*Zea mays* L.) inbred lines and hybrids under Khuzestan Province conditions. *J. Agric. Sci. Technol., 13*(1), 111-121.

Kotak, S., Larkindale, J., Lee, U., von Koskull-Döring, P., Vierling, E., Scharf, K.D. (2007). Complexity of the heat stress response in plants. *Curr. Opin. Plant Biol., 10*(3), 310-316.
[http://dx.doi.org/10.1016/j.pbi.2007.04.011] [PMID: 17482504]

Kumar, S., Gupta, D., Nayyar, H. (2012). Comparative response of maize and rice genotypes to heat stress: Status of oxidative stress and antioxidants. *Acta Physiol. Plant., 34*(1), 75-86.
[http://dx.doi.org/10.1007/s11738-011-0806-9]

Li, H.C., Zhang, H.N., Li, G.L., Liu, Z.H., Zhang, Y.M., Zhang, H.M., Guo, X.L. (2015). Expression of maize heat shock transcription factor gene ZmHsf06 enhances the thermotolerance and drought-stress tolerance of transgenic Arabidopsis. *Funct. Plant Biol., 42*(11), 1080-1091.
[http://dx.doi.org/10.1071/FP15080] [PMID: 32480747]

Li, G., Zhang, Y., Zhang, H., Zhang, Y., Zhao, L., Liu, Z. (2019). Characteristics and Regulating Role in Thermotolerance of the Heat Shock Transcription Factor ZmHsf12 from *Zea mays* L. *J. Plant Biol., 62*(5), 329-341.http://link.springer.com/10.1007/s12374-019-0067-5
[http://dx.doi.org/10.1007/s12374-019-0067-5]

liang, Li G, H ning, Zhang, H, Shao , G yan, Wang ( 2019). a new heat shock transcription factor from *Zea mays* L. improves thermotolerance in *Arabidopsis thaliana* and rescues thermotolerance defects of the athsfa2 mutant. *Plant Sci., 283*( December 2018), 84-375.
[http://dx.doi.org/https://doi.org/10.1016/j.plantsci.2019.03.002]

Li, Z.G., Zhu, L.P. (2015). Hydrogen sulfide donor sodium hydrosulfide-induced accumulation of betaine is involved in the acquisition of heat tolerance in maize seedlings. *Rev. Bras. Bot. Braz. J. Bot., 38*(1), 31-38.
[http://dx.doi.org/10.1007/s40415-014-0106-x]

Li, S., Zhou, X., Chen, L., Huang, W., Yu, D. (2010). Functional characterization of *Arabidopsis thaliana* WRKY39 in heat stress. *Mol. Cells, 29*(5), 475-483.

[http://dx.doi.org/10.1007/s10059-010-0059-2] [PMID: 20396965]

Liu, H.C., Liao, H.T., Charng, Y.Y. (2011). The role of class A1 heat shock factors (HSFA1s) in response to heat and other stresses in Arabidopsis. *Plant Cell Environ., 34*(5), 738-751.
[http://dx.doi.org/10.1111/j.1365-3040.2011.02278.x] [PMID: 21241330]

Liu, H.T., Gao, F., Li, G.L., Han, J.L., Liu, D.L., Sun, D.Y., Zhou, R.G. (2008). The calmodulin-binding protein kinase 3 is part of heat-shock signal transduction in *Arabidopsis thaliana. Plant J., 55*(5), 760-773.
[http://dx.doi.org/10.1111/j.1365-313X.2008.03544.x] [PMID: 18466301]

Li, P., Cao, W., Fang, H., Xu, S., Yin, S., Zhang, Y., Lin, D., Wang, J., Chen, Y., Xu, C., Yang, Z. (2017). Transcriptomic profiling of the maize (*Zea mays* L.) leaf response to abiotic stresses at the seedling stage. *Front. Plant Sci., 8*(March), 290.
[http://dx.doi.org/10.3389/fpls.2017.00290] [PMID: 28298920]

Li, Z., Srivastava, R., Tang, J., Zheng, Z., Howell, S.H. (2018). Cis-effects condition the induction of a major unfolded protein response factor, ZmbZIP60, in response to heat stress in Maize. *Front. Plant Sci., 9*(June), 833.
[http://dx.doi.org/10.3389/fpls.2018.00833] [PMID: 30008724]

Lizaso, J.I., Ruiz-Ramos, M., Rodríguez, L., Gabaldon-Leal, C., Oliveira, J.A., Lorite, I.J. (2017). Impact of high temperatures in maize: Phenology and yield components. *F Crop Res., 2018*(216), 129-140.
[http://dx.doi.org/10.1016/j.fcr.2017.11.013]

Lobell, D.B., Bänziger, M., Magorokosho, C., Vivek, B. (2011). Nonlinear heat effects on African maize as evidenced by historical yield trials. *Nat. Clim. Chang., 1*(1), 42-45.
[http://dx.doi.org/10.1038/nclimate1043]

Lu, C.M., Zhang, J.H. (2000). Photosystem II photochemistry and its sensitivity to heat stress in maize plants as affected by nitrogen deficiency. *J. Plant Physiol., 157*(1), 124-130.
[http://dx.doi.org/10.1016/S0176-1617(00)80145-5]

Machado, S., Paulsen, G.M. (2001). Combined effects of drought and high temperature on water relations of wheat and sorghum. *Plant Soil, 233*(2), 179-187.
[http://dx.doi.org/10.1023/A:1010346601643]

Mallikarjuna, M.G., Nepolean, T., Hossain, F., Manjaiah, K.M., Singh, A.M., Gupta, H.S. (2014). Genetic variability and correlation of kernel micronutrients among exotic quality protein maize inbreds and their utility in breeding programme. *Indian J. Genet. Plant Breed., 74*(2), 166-173.
[http://dx.doi.org/10.5958/0975-6906.2014.00152.7]

Mayer, L.I., Rattalino Edreira, J.I., Maddonni, G.A. (2014). Oil yield components of maize crops exposed to heat stress during early and late Grain-Filling stages. *Crop Sci., 54*(5), 2236-2250.
[http://dx.doi.org/10.2135/cropsci2013.11.0795]

Ma, H., Liu, C., Li, Z., Ran, Q., Xie, G., Wang, B., Fang, S., Chu, J., Zhang, J. (2018). ZmbZIP4 contributes to stress resistance in maize by regulating ABA synthesis and root development. *Plant Physiol., 178*(2), 753-770.
[http://dx.doi.org/10.1104/pp.18.00436] [PMID: 30126870]

McNellie, J.P., Chen, J., Li, X., Yu, J. (2018). Genetic mapping of foliar and tassel heat stress tolerance in maize. *Crop Sci., 58*(6), 2484-2493.
[http://dx.doi.org/10.2135/cropsci2018.05.0291]

Meuwissen, T., Goddard, M. (2010). Accurate prediction of genetic values for complex traits by whole-genome resequencing. *Genetics, 185*(2), 623-631. http://www.genetics.org/ cgi/ doi/ 10.1534/ genetics.110.116590
[http://dx.doi.org/10.1534/genetics.110.116590] [PMID: 20308278]

Meiri, D., Breiman, A. (2009). Arabidopsis ROF1 (FKBP62) modulates thermotolerance by interacting with HSP90.1 and affecting the accumulation of HsfA2-regulated sHSPs. *Plant J., 59*(3), 387-399.
[http://dx.doi.org/10.1111/j.1365-313X.2009.03878.x] [PMID: 19366428]

Manoj, K., Ghimire, S.K., Ojha, B.R., Shrestha, J. (2018). Genetic diversity for heat tolerant related traits in maize inbred lines. *Agricultura., 105*(1–2), 23-34.

Mishkind, M., Vermeer, J.E.M., Darwish, E., Munnik, T. (2009). Heat stress activates phospholipase D and triggers PIP accumulation at the plasma membrane and nucleus. *Plant J., 60*(1), 10-21.
[http://dx.doi.org/10.1111/j.1365-313X.2009.03933.x] [PMID: 19500308]

Mittler, R., Finka, A., Goloubinoff, P. (2012). How do plants feel the heat? *Trends Biochem. Sci., 37*(3), 118-125.
[http://dx.doi.org/10.1016/j.tibs.2011.11.007] [PMID: 22236506]

Mittal, S., Mallikarjuna, M.G., Rao, A.R., Jain, P.A., Dash, P.K., Thirunavukkarasu, N. (2017). Comparative analysis of CDPK family in maize, Arabidopsis, rice, and sorghum revealed potential targets for drought tolerance improvement. *Front Chem., 5*(DEC), 115.
[http://dx.doi.org/10.3389/fchem.2017.00115] [PMID: 29312925]

Mohamed, H.I., Ashry, N.A., Ghonaim, M.M. (2019). Physiological and biochemical effects of heat shock stress and determination of molecular markers related to heat tolerance in maize hybrids. *Gesunde Pflanz, 71*(3), 213-222.

Miroshnichenko, S., Tripp, J., Nieden, Uz., Neumann, D., Conrad, U., Manteuffel, R. (2005). Immunomodulation of function of small heat shock proteins prevents their assembly into heat stress granules and results in cell death at sublethal temperatures. *Plant J., 41*(2), 269-281.
[http://dx.doi.org/10.1111/j.1365-313X.2004.02290.x] [PMID: 15634203]

Momcilovic, I., Ristic, Z. (2004). Localization and abundance of chloroplast protein synthesis elongation factor (EF-Tu) and heat stability of chloroplast stromal proteins in maize. *Plant Sci., 166*(1), 81-88.
[http://dx.doi.org/10.1016/j.plantsci.2003.08.009]

Morimoto, R.I. (1998). Regulation of the heat shock transcriptional response: cross talk between a family of heat shock factors, molecular chaperones, and negative regulators. *Genes Dev., 12*(24), 3788-3796.
[http://dx.doi.org/10.1101/gad.12.24.3788] [PMID: 9869631]

Momcilovic, I., Ristic, Z. (2007). Expression of chloroplast protein synthesis elongation factor, EF-Tu, in two lines of maize with contrasting tolerance to heat stress during early stages of plant development. *J. Plant Physiol., 164*(1), 90-99.
[http://dx.doi.org/10.1016/j.jplph.2006.01.010] [PMID: 16542752]

Mallikarjuna, M.G., Thirunavukkarasu, N., Hossain, F., Bhat, J.S., Jha, S.K., Rathore, A., Agrawal, P.K., Pattanayak, A., Reddy, S.S., Gularia, S.K., Singh, A.M., Manjaiah, K.M., Gupta, H.S. (2015). Stability performance of inductively coupled plasma mass spectrometry-phenotyped kernel minerals concentration and grain yield in maize in different agro-climatic zones. *PLoS One, 10*(9), e0139067.
[http://dx.doi.org/10.1371/journal.pone.0139067] [PMID: 26406470]

Meseka, S., Menkir, A., Bossey, B., Mengesha, W. (2018). Performance assessment of drought tolerant maize hybrids under combined drought and heat stress. *Agronomy (Basel), 8*(12).
[http://dx.doi.org/10.3390/agronomy8120274]

Mickelbart, M.V., Hasegawa, P.M., Bailey-Serres, J. (2015). Genetic mechanisms of abiotic stress tolerance that translate to crop yield stability. *Nat. Rev. Genet., 16*(4), 237-251.
[http://dx.doi.org/10.1038/nrg3901] [PMID: 25752530]

Murari, L., Nozawa, T., Emori, S., Harasawa, H., Takahashi, K., Kimoto, M. (2001). Future climate change: Implications for Indian summer monsoon and its variability. *Curr. Sci., 81*(9), 1196-1207.

Naveed, M., Ahsan, M., Akram, H.M., Aslam, M., Ahmed, N. (2016). Measurement of cell membrane thermo-stability and leaf temperature for heat tolerance in maize (*Zea mays* L): Genotypic variability and inheritance pattern. *Maydica, 61*(2)

Naveed, M., Ahsan, M., Akram, H.M., Aslam, M., Ahmed, N. (2016). Genetic effects conferring heat tolerance in a cross of tolerant × susceptible maize (*Zea mays* L.) genotypes. *Front. Plant Sci., 7*(June), 729.http://journal.frontiersin.org/Article/10.3389/fpls.2016.00729/abstract

[http://dx.doi.org/10.3389/fpls.2016.00729] [PMID: 27313583]

Nieto-Sotelo, J., Martínez, L.M., Ponce, G., Cassab, G.I., Alagón, A., Meeley, R.B., Ribaut, J.M., Yang, R. (2002). Maize HSP101 plays important roles in both induced and basal thermotolerance and primary root growth. *Plant Cell, 14*(7), 1621-1633.
[http://dx.doi.org/10.1105/tpc.010487] [PMID: 12119379]

Nelimor, C., Badu-Apraku, B., Tetteh, A.Y., N'guetta, A.S.P. (2019). Assessment of genetic diversity for drought, heat and combined drought and heat stress tolerance in early maturing maize landraces. *Plants, 8*(11), 518.https://www.mdpi.com/2223-7747/8/11/518
[http://dx.doi.org/10.3390/plants8110518] [PMID: 31744251]

Noor, J.J., Vinayan, M.T., Umar, S., Devi, P., Iqbal, M., Seetharam, K. (2019). Morpho-physiological traits associated with heat stress tolerance in tropical maize (*Zea mays* L.) at the reproductive stage. *Aust. J. Crop Sci., 13*(4), 536-545.
[http://dx.doi.org/10.21475/ajcs.19.13.04.p1448]

Noshay, J.M., Anderson, S.N., Zhou, P., Ji, L., Ricci, W., Lu, Z., Stitzer, M.C., Crisp, P.A., Hirsch, C.N., Zhang, X., Schmitz, R.J., Springer, N.M. (2019). Monitoring the interplay between transposable element families and DNA methylation in maize. *PLoS Genet., 15*(9), e1008291.
[http://dx.doi.org/10.1371/journal.pgen.1008291] [PMID: 31498837]

Noshay, J.M., Crisp, P.A., Springer, N.M. (2018). The Maize Methylome.
[http://dx.doi.org/10.1007/978-3-319-97427-9_6]

Obata, T., Witt, S., Lisec, J., Palacios-Rojas, N., Florez-Sarasa, I., Yousfi, S., Araus, J.L., Cairns, J.E., Fernie, A.R. (2015). Metabolite profiles of maize leaves in drought, heat, and combined stress field trials reveal the relationship between metabolism and grain yield. *Plant Physiol., 169*(4), 2665-2683.
[http://dx.doi.org/10.1104/pp.15.01164] [PMID: 26424159]

Oliver, S.C., Venis, M.A., Freedman, R.B., Napier, R.M. (1995). Regulation of synthesis and turnover of maize auxin-binding protein and observations on its passage to the plasma membrane: comparisons to maize immunoglobulin-binding protein cognate. *Planta, 197*(3), 465-474.
[http://dx.doi.org/10.1007/BF00196668] [PMID: 8580760]

Pegoraro, C., Mertz, L.M., da Maia, L.C., Rombaldi, C.V., de Oliveira, A.C. (2011). Importance of heat shock proteins in maize. *J. Crop Sci. Biotechnol., 14*(2), 85-95.
[http://dx.doi.org/10.1007/s12892-010-0119-3]

Petolino, J.F., Cowen, N.M., Thompson, S.A., Mitchell, J.C. (1990). Gamete selection for heat-stress tolerance in maize. *J. Plant Physiol., 136*(2), 219-224.
[http://dx.doi.org/10.1016/S0176-1617(11)81669-X]

Petolino, J.F., Cowen, N.M., Thompson, S.A., Mitchell, J.C. (1992). Gamete selection for heat stress tolerance in maize. *Angiosperm Pollen and Ovules.* (pp. 355-358). New York, NY: Springer New York.http://link.springer.com/10.1007/978-1-4612-2958-2_57
[http://dx.doi.org/10.1007/978-1-4612-2958-2_57]

Peng, S., Huang, J., Sheehy, J.E., Laza, R.C., Visperas, R.M., Zhong, X., Centeno, G.S., Khush, G.S., Cassman, K.G. (2004). Rice yields decline with higher night temperature from global warming. *Proc. Natl. Acad. Sci. USA, 101*(27), 9971-9975.
[http://dx.doi.org/10.1073/pnas.0403720101] [PMID: 15226500]

Pingali, P. (2001). Meeting world maize needs: technological opportunities and priorities for the public sector. *CIMMYT 1999–2000 World Maize Facts and Trends.*

Prasanna, B.M. (2012). Diversity in global maize germplasm: characterization and utilization. *J. Biosci., 37*(5), 843-855.
[http://dx.doi.org/10.1007/s12038-012-9227-1] [PMID: 23107920]

Qian, Y., Ren, Q., Zhang, J., Chen, L. (2019). Transcriptomic analysis of the maize (*Zea mays* L.) inbred line B73 response to heat stress at the seedling stage. *Gene, 692*(692), 68-78.

[http://dx.doi.org/10.1016/j.gene.2018.12.062] [PMID: 30641208]

Qian, Y., Hu, W., Liao, J., Zhang, J., Ren, Q. (2019). The Dynamics of DNA methylation in the maize (*Zea mays* L.) inbred line B73 response to heat stress at the seedling stage. *Biochem. Biophys. Res. Commun., 512*(4), 742-749.
[http://dx.doi.org/10.1016/j.bbrc.2019.03.150] [PMID: 30926168]

Qu, A.L., Ding, Y.F., Jiang, Q., Zhu, C. (2013). Molecular mechanisms of the plant heat stress response. *Biochem. Biophys. Res. Commun., 432*(2), 203-207.
[http://dx.doi.org/10.1016/j.bbrc.2013.01.104] [PMID: 23395681]

Qu, M., Chen, G., Bunce, J.A., Zhu, X., Sicher, R.C. (2018). Systematic biology analysis on photosynthetic carbon metabolism of maize leaf following sudden heat shock under elevated $CO_2$. *Sci. Rep., 8*(1), 7849.
[http://dx.doi.org/10.1038/s41598-018-26283-x] [PMID: 29777170]

Rodríguez, M., Canales, E., Borras-Hidalgo, O. (2005). Molecular aspects of osmotic stress in plants. *Biotecnol. Apl., 22*(1), 1-10.

Rao, D., Momcilovic, I., Kobayashi, S., Callegari, E., Ristic, Z. (2004). Chaperone activity of recombinant maize chloroplast protein synthesis elongation factor, EF-Tu. *Eur. J. Biochem., 271*(18), 3684-3692.
[http://dx.doi.org/10.1111/j.1432-1033.2004.04309.x] [PMID: 15355346]

Rani, N., Ni, R.B.P., Kumari, S., Singh, B., Kumari, H. (2018). Genetic diversity under heat stress condition in maize (*Zea mays* L.) inbred lines. *Int. J. Curr. Microbiol. Appl. Sci., 7*(1), 4539-4547.

Rahman, S., Arif, M., Hussain, K., Hussain, S., Mukhtar, T., Razaq, A. (2013). Evaluation of maize hybrids for tolerance to high temperature stress in central Punjab. *Am. J. Bioeng. Biotechnol., 1*(1), 30-36.
[http://dx.doi.org/10.7726/ajbebt.2013.1003]

Rattalino Edreira, J.I., Otegui, M.E. (2012). Heat stress in temperate and tropical maize hybrids: Differences in crop growth, biomass partitioning and reserves use. *F Crop Res., 130*, 87-98.
[http://dx.doi.org/10.1016/j.fcr.2012.02.009]

Raju, BM., Rama Rao, CA, Rao, K V., Srinivasarao, C, Samuel, J (2018). Assessing unrealized yield potential of maize producing districts in India. *Curr. Sci., 114*(9), 1885-1894.
[http://dx.doi.org/10.18520/cs/v114/i09/1885-1893]

Reddy, A.S.N., Ali, G.S., Celesnik, H., Day, I.S. (2011). Coping with stresses: Roles of calcium- and calcium/calmodulin-regulated gene expression. *Plant Cell, 23*(6), 2010-2032.
[http://dx.doi.org/10.1105/tpc.111.084988] [PMID: 21642548]

Reed, A.J., Singletary, G.W. (1989). Roles of carbohydrate supply and phytohormones in maize kernel abortion. *Plant Physiol., 91*(3), 986-992.
[http://dx.doi.org/10.1104/pp.91.3.986] [PMID: 16667166]

Ristic, Z., Wilson, K., Nelsen, C., Momcilovic, I., Kobayashi, S., Meeley, R. (2004). A maize mutant with decreased capacity to accumulate chloroplast protein synthesis elongation factor (EF-Tu) displays reduced tolerance to heat stress. *Plant Sci., 167*(6), 1367-1374.
[http://dx.doi.org/10.1016/j.plantsci.2004.07.016]

Rosegrant, M.W., Ringler, C., Zhu, T. (2009). Water for Agriculture: Maintaining Food Security under Growing Scarcity. *Annu. Rev. Environ. Resour., 34*(1), 205-222.
[http://dx.doi.org/10.1146/annurev.environ.030308.090351]

Schlenker, W., Roberts, M.J. (2009). Nonlinear temperature effects indicate severe damages to U.S. crop yields under climate change. *Proc. Natl. Acad. Sci. USA, 106*(37), 15594-15598.
[http://dx.doi.org/10.1073/pnas.0906865106] [PMID: 19717432]

Shiferaw, B., Prasanna, B.M., Hellin, J., Bänziger, M. (2011). Crops that feed the world 6. Past successes and future challenges to the role played by maize in global food security. *Food Secur., 3*(3), 307-327.
[http://dx.doi.org/10.1007/s12571-011-0140-5]

Sunoj, V.S.J., Shroyer, K.J., Jagadish, S.V.K., Prasad, P.V.V. (2016). Diurnal temperature amplitude alters

physiological and growth response of maize (*Zea mays* L.) during the vegetative stage. *Environ. Exp. Bot.*, *130*, 113-121.
[http://dx.doi.org/10.1016/j.envexpbot.2016.04.007]

Schoper, J.B., Lambert, R.J., Vasilas, B.L. (1987). Pollen viability, pollen shedding, and combining ability for tassel heat tolerance in maize. *Crop Sci.*, *27*(1), 27.
[http://dx.doi.org/10.2135/cropsci1987.0011183X002700010007x]

Singletary, G.W., Banisadr, R., Keeling, P.L. (1994). Heat stress during grain filling in maize : Effects on carbohydrate storage and metabolism. *Aust. J. Plant Physiol.*, *21*, 829-841.

Savchenko, G.E., Klyuchareva, E.A., Abramchik, L.M., Serdyuchenko, E.V. (2002). Effect of periodic heat shock on the inner membrane system of etioplasts. *Russ. J. Plant Physiol.*, *49*(3), 349-359.
[http://dx.doi.org/10.1023/A:1015592902659]

Sejima, T., Takagi, D., Fukayama, H., Makino, A., Miyake, C. (2014). Repetitive short-pulse light mainly inactivates photosystem I in sunflower leaves. *Plant Cell Physiol.*, *55*(6), 1184-1193.
[http://dx.doi.org/10.1093/pcp/pcu061] [PMID: 24793753]

Sehgal, A., Sita, K., Siddique, K.H.M., Kumar, R., Bhogireddy, S., Varshney, R.K., HanumanthaRao, B., Nair, R.M., Prasad, P.V.V., Nayyar, H. (2018). Drought or/and heat-stress effects on seed filling in food crops: Impacts on functional biochemistry, seed yields, and nutritional quality. *Front. Plant Sci.*, *9*(November), 1705.
[http://dx.doi.org/10.3389/fpls.2018.01705] [PMID: 30542357]

Sanmiya, K., Suzuki, K., Egawa, Y., Shono, M. (2004). Mitochondrial small heat-shock protein enhances thermotolerance in tobacco plants. *FEBS Lett.*, *557*(1-3), 265-268.
[http://dx.doi.org/10.1016/S0014-5793(03)01494-7] [PMID: 14741379]

Swindell, W.R., Huebner, M., Weber, A.P. (2007). Transcriptional profiling of Arabidopsis heat shock proteins and transcription factors reveals extensive overlap between heat and non-heat stress response pathways. *BMC Genomics*, *8*, 125.
[http://dx.doi.org/10.1186/1471-2164-8-125] [PMID: 17519032]

Saidi, Y., Finka, A., Goloubinoff, P. (2011). Heat perception and signalling in plants: a tortuous path to thermotolerance. *New Phytol.*, *190*(3), 556-565.
[http://dx.doi.org/10.1111/j.1469-8137.2010.03571.x] [PMID: 21138439]

Saidi, Y., Peter, M., Finka, A., Cicekli, C., Vigh, L., Goloubinoff, P. (2010). Membrane lipid composition affects plant heat sensing and modulates Ca(2+)-dependent heat shock response. *Plant Signal. Behav.*, *5*(12), 1530-1533.
[http://dx.doi.org/10.4161/psb.5.12.13163] [PMID: 21139423]

Saidi, Y., Finka, A., Muriset, M., Bromberg, Z., Weiss, Y.G., Maathuis, F.J.M., Goloubinoff, P. (2009). The heat shock response in moss plants is regulated by specific calcium-permeable channels in the plasma membrane. *Plant Cell*, *21*(9), 2829-2843.
[http://dx.doi.org/10.1105/tpc.108.065318] [PMID: 19773386]

Saidi, Y., Finka, A., Chakhporanian, M., Zrÿd, J.P., Schaefer, D.G., Goloubinoff, P. (2005). Controlled expression of recombinant proteins in Physcomitrella patens by a conditional heat-shock promoter: a tool for plant research and biotechnology. *Plant Mol. Biol.*, *59*(5), 697-711.
[http://dx.doi.org/10.1007/s11103-005-0889-z] [PMID: 16270224]

Suri, S.S., Dhindsa, R.S. (2008). A heat-activated MAP kinase (HAMK) as a mediator of heat shock response in tobacco cells. *Plant Cell Environ.*, *31*(2), 218-226.
[http://dx.doi.org/10.1111/j.1365-3040.2007.01754.x] [PMID: 17996015]

Sanders, D., Brownlee, C., Harper, J.F. (1999). Communicating with calcium. *Plant Cell*, *11*(4), 691-706.
[http://dx.doi.org/10.1105/tpc.11.4.691] [PMID: 10213787]

Suzuki, N., Sejima, H., Tam, R., Schlauch, K., Mittler, R. (2011). Identification of the MBF1 heat-response regulon of *Arabidopsis thaliana*. *Plant J.*, *66*(5), 844-851.

[http://dx.doi.org/10.1111/j.1365-313X.2011.04550.x] [PMID: 21457365]

Suzuki, N., Miller, G., Morales, J., Shulaev, V., Torres, M.A., Mittler, R. (2011). Respiratory burst oxidases: the engines of ROS signaling. *Curr. Opin. Plant Biol., 14*(6), 691-699.
[http://dx.doi.org/10.1016/j.pbi.2011.07.014] [PMID: 21862390]

Shi, J., Yan, B., Lou, X., Ma, H., Ruan, S. (2017). Comparative transcriptome analysis reveals the transcriptional alterations in heat-resistant and heat-sensitive sweet maize (*Zea mays* L.) varieties under heat stress. *BMC Plant Biol., 17*(1), 26.
[http://dx.doi.org/10.1186/s12870-017-0973-y] [PMID: 28122503]

Sun, C.X., Li, M.Q., Gao, X.X., Liu, L.N., Wu, X.F., Zhou, J.H. (2016). Metabolic response of maize plants to multi-factorial abiotic stresses. *Plant Biol, 18* (Suppl. 1), 120-129.
[http://dx.doi.org/10.1111/plb.12305] [PMID: 25622534]

Sun, C.X., Gao, X.X., Li, M.Q., Fu, J.Q., Zhang, Y.L. (2016). Plastic responses in the metabolome and functional traits of maize plants to temperature variations. *Plant Biol (Stuttg), 18*(2), 249-261.
[http://dx.doi.org/10.1111/plb.12378] [PMID: 26280133]

Shikha, M., Kanika, A., Rao, A.R., Mallikarjuna, M.G., Gupta, H.S., Nepolean, T. (2017). Genomic selection for drought tolerance using genome-wide snapshots in maize. *Front. Plant Sci., 8*, 550.
[http://dx.doi.org/10.3389/fpls.2017.00550] [PMID: 28484471]

Sun, L., Liu, Y., Kong, X., Zhang, D., Pan, J., Zhou, Y., Wang, L., Li, D., Yang, X. (2012). ZmHSP16.9, a cytosolic class I small heat shock protein in maize (*Zea mays*), confers heat tolerance in transgenic tobacco. *Plant Cell Rep., 31*(8), 1473-1484.
[http://dx.doi.org/10.1007/s00299-012-1262-8] [PMID: 22534681]

Settles, A.M., Ribeiro, C. *Mitigation of maize heat stress with recombinant 6-phosphogluconate dehydrogenase. United States of America: World Intellectual Property Organization,* WO 2019/090265 Al. (2019). 1-91.

Shi, J., Gao, H., Wang, H., Lafitte, H.R., Archibald, R.L., Yang, M., Hakimi, S.M., Mo, H., Habben, J.E. (2017). ARGOS8 variants generated by CRISPR-Cas9 improve maize grain yield under field drought stress conditions. *Plant Biotechnol. J., 15*(2), 207-216.
[http://dx.doi.org/10.1111/pbi.12603] [PMID: 27442592]

Tack, J., Lingenfelser, J., Jagadish, S.V.K. (2017). Disaggregating sorghum yield reductions under warming scenarios exposes narrow genetic diversity in US breeding programs. *Proc. Natl. Acad. Sci. USA, 114*(35), 9296-9301.
[http://dx.doi.org/10.1073/pnas.1706383114] [PMID: 28808013]

Tao, Z.Q., Chen, Y.Q., Chao, L.I., Zou, J.X., Peng, Y.A.N., Yuan, S.F., Xia, W.U., Peng, S.U.I. ( 2016). The causes and impacts for heat stress in spring maize during grain filling in the North China Plain — A review. *J Integr Agric., 15*( 12), 2677.
[http://dx.doi.org/10.1016/S2095-3119(16)61409-0]

Tao, Z., Chen, Y., Li, C., Yuan, S., Shi, J., Gao, W., Sui, P. (2013). Path analysis between yield of spring maize and meteorological factors at different sowing times in North China Low Plain. *Acta Agronomica Sinica, 39*(9), 1628-34.
[http://dx.doi.org/10.3724/SP.J.1006.2013.01628]

Tandzi, L.N., Bradley, G., Mutengwa, C. (2019). Morphological responses of maize to drought, heat and combined stresses at seedling stage. *J. Biol. Sci., 19*(1), 7-16.
[http://dx.doi.org/10.3923/jbs.2019.7.16]

Tigchelaar, M., Battisti, D.S., Naylor, R.L., Ray, D.K. (2018). Future warming increases probability of globally synchronized maize production shocks. *Proc. Natl. Acad. Sci. USA, 115*(26), 6644-6649.
[http://dx.doi.org/10.1073/pnas.1718031115] [PMID: 29891651]

Tesfaye, K., Kruseman, G., Cairns, J.E., Zaman-Allah, M., Wegary, D., Zaidi, P.H. (2017). Potential benefits of drought and heat tolerance for adapting maize to climate change in tropical environments. *Clim. Risk*

*Manage., 2018*(19), 106-119.
[http://dx.doi.org/10.1016/j.crm.2017.10.001]

Thirunavukkarasu, N, Sharma, R, Singh, N, Shiriga, K, Mohan, S, Mittal, S (2017). *Genomewide expression and functional interactions of genes under drought stress in maize.*
[http://dx.doi.org/10.1155/2017/2568706]

Tiwari, Y.K., Yadav, S.K. (2019). High temperature stress tolerance in maize (*Zea mays* L.): Physiological and molecular mechanisms. *J. Plant Biol., 62*(2), 93-102.
[http://dx.doi.org/10.1007/s12374-018-0350-x]

Tripp, J., Mishra, S.K., Scharf, K.D. (2009). Functional dissection of the cytosolic chaperone network in tomato mesophyll protoplasts. *Plant Cell Environ., 32*(2), 123-133.
[http://dx.doi.org/10.1111/j.1365-3040.2008.01902.x] [PMID: 19154229]

Trachsel, S., Dhliwayo, T., Gonzalez Perez, L., Mendoza Lugo, J.A., Trachsel, M. (2019). Estimation of physiological genomic estimated breeding values (PGEBV) combining full hyperspectral and marker data across environments for grain yield under combined heat and drought stress in tropical maize (*Zea mays* L.). *PLoS One, 14*(3), e0212200.
[http://dx.doi.org/10.1371/journal.pone.0212200] [PMID: 30893307]

Van Inghelandt, D., Frey, F.P., Ries, D., Stich, B. (2019). QTL mapping and genome-wide prediction of heat tolerance in multiple connected populations of temperate maize. *Sci. Rep., 9*(1), 1-16.
[http://dx.doi.org/10.1038/s41598-019-50853-2] [PMID: 30626917]

Wang, Q.L., Chen, J.H., He, N.Y., Guo, F.Q. (2018). Metabolic reprogramming in chloroplasts under heat stress in plants. *Int. J. Mol. Sci., 19*(3), 9-11.
[http://dx.doi.org/10.3390/ijms19030849] [PMID: 29538307]

Wang, X., Xu, C., Cai, X., Wang, Q., Dai, S. (2017). Heat-responsive photosynthetic and signaling pathways in plants: Insight from proteomics. *Int. J. Mol. Sci., 18*(10), E2191.
[http://dx.doi.org/10.3390/ijms18102191] [PMID: 29053587]

Wang, C.T., Ru, J.N., Liu, Y.W., Li, M., Zhao, D., Yang, J.F., Fu, J.D., Xu, Z.S. (2018). Maize WRKY transcription factor zmwrky106 confers drought and heat tolerance in transgenic plants. *Int. J. Mol. Sci., 19*(10), 1-15.
[http://dx.doi.org/10.3390/ijms19103046] [PMID: 30301220]

Wahid, A., Gelani, S., Ashraf, M., Foolad, M.R. (2007). Heat tolerance in plants: An overview. *Environ. Exp. Bot., 61*(3), 199-223.
[http://dx.doi.org/10.1016/j.envexpbot.2007.05.011]

Wilhelm, E.P., Mullen, R.E., Keeling, P.L., Singletary, G.W. (1999). Heat stress during grain filling in maize: Effects on kernel growth and metabolism. *Crop Sci., 39*(6), 1733-1741.
[http://dx.doi.org/10.2135/cropsci1999.3961733x]

Xu, S., Li, J., Zhang, X., Wei, H., Cui, L. (2006). Effects of heat acclimation pretreatment on changes of membrane lipid peroxidation, antioxidant metabolites, and ultrastructure of chloroplasts in two cool-season turfgrass species under heat stress. *Environ. Exp. Bot., 56*(3), 274-285.
[http://dx.doi.org/10.1016/j.envexpbot.2005.03.002]

Yamamoto, K., Sakamoto, H., Momonoki, Y.S. (2011). Maize acetylcholinesterase is a positive regulator of heat tolerance in plants. *J. Plant Physiol., 168*(16), 1987-1992.
[http://dx.doi.org/10.1016/j.jplph.2011.06.001] [PMID: 21757255]

Yang, Z., Sinclair, T.R., Zhu, M., Messina, C.D., Cooper, M., Hammer, G.L. (2012). Temperature effect on transpiration response of maize plants to vapour pressure deficit. *Environ. Exp. Bot., 78*, 157-162.
[http://dx.doi.org/10.1016/j.envexpbot.2011.12.034]

Young, T.E., Ling, J., Geisler-Lee, C.J., Tanguay, R.L., Caldwell, C., Gallie, D.R. (2001). Developmental and thermal regulation of the maize heat shock protein, HSP101. *Plant Physiol., 127*(3), 777-791.
[http://dx.doi.org/10.1104/pp.010160] [PMID: 11706162]

Yong, W.S., Hsu, F.M., Chen, P.Y. (2016). Profiling genome-wide DNA methylation. *Epigenetics Chromatin,   9*(1), 26.
[http://dx.doi.org/10.1186/s13072-016-0075-3] [PMID: 27358654]

Yousaf, M.I., Hussain, K., Hussain, S., Ghani, A., Arshad, M., Mumtaz, A. (2018). Characterization of indigenous and exotic maize hybrids for grain yield and quality traits under heat stress. *Int. J. Agric. Biol., 20*(2), 333-337.
[http://dx.doi.org/10.17957/IJAB/15.0493]

Yu, S.M., Lo, S.F., Ho, T.D. (2015). Source-sink communication: regulated by hormone, nutrient, and stress cross-signaling. *Trends Plant Sci.,   20*(12), 844-857.
[http://dx.doi.org/10.1016/j.tplants.2015.10.009] [PMID: 26603980]

Yuan, Y., Cairns, J.E., Babu, R., Gowda, M., Makumbi, D., Magorokosho, C., Zhang, A., Liu, Y., Wang, N., Hao, Z., San Vicente, F., Olsen, M.S., Prasanna, B.M., Lu, Y., Zhang, X. (2019). Genome-wide association mapping and genomic prediction analyses reveal the genetic architecture of grain yield and flowering time under drought and heat stress conditions in maize. *Front. Plant Sci.,   9*(January), 1919.
[http://dx.doi.org/10.3389/fpls.2018.01919] [PMID: 30761177]

Zaidi, P.H., Zaman-Allah, M., Trachsel, S., Seetharam, K., Cairns, J.E., Vinayan, M.T. (2016). *Phenotyping for Abiotic Stress Tolerance in Maize : Heat Stress. A field manual..* CIMMYT.

Zhao, F, Zhang, D, Zhao, Y, Wang, W, Yang, H, Ta, F (2016). *The difference of physiological and proteomic changes in maize leaves adaptation to drought, heat, and combined both stresses.*
[http://dx.doi.org/10.3389/fpls.2016.01471]

Zhao, Y., Hu, F., Zhang, X., Wei, Q., Dong, J., Bo, C., Cheng, B., Ma, Q. (2019). Comparative transcriptome analysis reveals important roles of nonadditive genes in maize hybrid An'nong 591 under heat stress. *BMC Plant Biol.,   19*(1), 273.
[http://dx.doi.org/10.1186/s12870-019-1878-8] [PMID: 31234785]

Zhao, L., Li, C., Liu, T., Wang, X., Seng, S. (2012). Effect of high temperature during flowering on photosynthetic characteristics and grain yield and quality of different genotypes of maize (*Zea mays* L.). *Scientia Agri Sinica,   45*(23), 4947-4958.

Zheng, S.Z., Liu, Y.L., Li, B., Shang, Z.L., Zhou, R.G., Sun, D.Y. (2012). Phosphoinositide-specific phospholipase C9 is involved in the thermotolerance of Arabidopsis. *Plant J.,   69*(4), 689-700.
[http://dx.doi.org/10.1111/j.1365-313X.2011.04823.x] [PMID: 22007900]

Zhu, X.C., Song, F. (2011). Bin, Liu SQ, Liu TD. Effects of arbuscular mycorrhizal fungus on photosynthesis and water status of maize under high temperature stress. *Plant Soil,   346*(1), 189-199.
[http://dx.doi.org/10.1007/s11104-011-0809-8]

# CHAPTER 3

# Breeding Pearl Millet for Heat Stress Tolerance

**P. Sanjana Reddy**[*]

*ICAR-Indian Institute of Millets Research, Rajendranagar, Hyderabad 500030, India*

**Abstract:** Pearl millet (*Pennisetum glaucum* L. R. Br.) is an important cereal crop grown by resource poor farmers of semi-arid and arid tropics. It is grown for food in Asia, Africa and Latin America and for fodder in the USA, Australia and Brazil. Due to its inherent ability for tolerance to temperature and drought, salinity and nutrient-poor soils, it is grown in harsh environments. Owing to the changing climatic conditions, the crop holds promise for food and nutrition security for the increasing world population. However, high temperature stress is one of the main reasons for low productivity in pearl millet under semi-arid and arid environments. Further improvement for thermo tolerance is needed for the economization of agriculture.

Heat stress (HS) is a complex function of intensity, duration, and rate of increase in temperature. Tolerance mechanisms to HS are exhibited in all stages of crops such as seedling emergence, vegetative stage, flowering/ reproductive, and grain filling stages. For surviving under HS, crop plants show short-term (avoidance) and long-term (adaptation) strategies. A wide range of plant developmental and physiological processes are negatively affected by HS. Heat tolerance (HT) has been linked to increased tolerance of the photosynthetic apparatus and correlated with increased capacity of scavenging and detoxifying of reactive oxygen species (ROS). Induction of thermotolerance may be ascribed to the maintenance of a better membrane thermostability (MTS) and low ROS accumulation due to improved antioxidant capacity, osmo-regulation of solutes and synthesis of heat shock proteins (HSPs). Heat tolerance can be evaluated by a field screening, lab cum field screening or laboratory screening protocols and testing under hotspot locations. In pearl millet, the studies on heat tolerance are limited and the few studies made to date suggest seedling thermo-tolerance index (STI), seed to seedling thermo-tolerance index (SSTI) in pearl millet, and heat tolerance index (HTI) are indicative of heat tolerance. However, these are not indicative of maturity stage traits wherein membrane thermo stability holds promise. For breeding for heat tolerance, information on genetic variability, gene action (additive and non-additive), heritability, stability and correlation are available. Landraces are adapted to their native environment and could be the potential sources of HT. Gene interaction on heat tolerance showed its complex nature of inheritance. Plants are relatively more sensitive to HT during reproductive than vegetative stages. Breeders should consider and devise tools for heat tolerance screening which directly links to the productivity of a crop. Different breeding strategies such as conventional

[*] **Corresponding author P. Sanjana Reddy:** ICAR-Indian Institute of Millets Research, Rajendranagar, Hyderabad 500030, India; E-mail: sanjana@millets.res.in

**Uday C. Jha, Harsh Nayyar and Sanjeev Gupta (Eds.)**

breeding methods, physiological trait-based breeding, molecular or transgenic approach can be applied individually or in combination for genetic improvement for heat tolerance.

**Keywords:** Breeding strategies, Growth stage, Gene action, Membrane thermostability, Screening, Thermo-tolerance index.

## INTRODUCTION

The arid and semiarid zones constitute about 41% of the land area of the world that are characterized by unpredictable and challenging environmental conditions that are detrimental to the optimum crop production (Safriel *et al.,* 2005). For every degree centigrade increase in average growing season temperature, the crop yields are estimated to reduce up to 17% (Lobell and Asner, 2003). Due to global climatic changes, a 1-4°C on an average is expected to increase by the end of the 21st century (Driedonks *et al.,* 2016). The climate change in terms of amount and distribution of rainfall and rise in temperatures will be detrimental to crop yield. Crops that are resilient to adverse climatic conditions would play an important role in sustainable food availability to the ever-increasing world population. Pearl millet is grown in hot semi-arid areas where it is better adapted than other crops and thereby has a great potential as an excellent genomic resource for isolation of candidate genes for tolerance to drought and heat stresses. Among the millets grown in India, 75% of the total area is occupied with pearl millet. During 2017-18, the crop was grown on about 7.4 million ha with an average production of 9.13 million tons and a yield of 1237 kg/ha (Directorate of Millets Development, 2019). Though pearl millet is a climate resilient crop, owing to its cultivation in harsh conditions, the abiotic stresses such as the low and erratic distribution of rainfall, high temperatures especially during seed germination, poor soil fertility limit the crop production to subsistence level. These abiotic stresses force the crop to complete its lifecycle thereby reducing the length of the growing period (LGP) thereby resulting in a reduction in productivity (Cooper *et al.,*2009).

Heat stress tolerance is the ability of the plant to evade the negative impact of heat stress and attain economic yields near to that of normal conditions (Wahid *et al.,* 2007). Tolerance varies from species to species and between genotypes within a species. At the cellular level, high temperatures trigger certain genes and production of metabolites which enhance the plant's ability to tolerate heat stress (Hasanuzzaman *et al.,* 2013). Understanding the response of plant leaf tissues to climate change would be helpful in order to be able to predict plant performance under various stress factors. The development of heat-tolerant hybrids/varieties and the generation of improved pre-breeding materials for any breeding program is crucial in meeting food security (Ortiz *et al.,* 2008).

# HEAT STRESS AT DIFFERENT PHENOLOGICAL STAGES

The effect of heat stress depends upon the phenological stage of the crop exposed to heat stress. The impact on the grain yield is more when these critical stages are exposed to heat stress. In pearl millet, the seedling stage is more vulnerable in areas where it encounters high soil temperatures during germination (Howarth *et al.,* 1997). Heat stress during flowering and seed development also affect the grain yield (Hall, 1992) especially in areas where pearl millet is grown during the summer season.

*Seed Germination:* Soil temperature determines both germination percentage and rate of germination. The percentage of final germination, rate of germination, seedling survival and growth increased with an increase in temperature (Pearson, 1975) and at around 42°C, the percentage and rate of germination decreased. Millet germination rate and plumule emergence had optima at 37–38°C (Ashraf and Hafeez, 2004). Seed size and density also affected seedling thermo tolerance (Gardner and Vanderlip, 1989). Small seed affects adequate plant stand and establishment of pearl millet in dry areas. Seedling emergence increased from 40% with small and low-density seed, to 62% with large and high-density seed. Germination, seedling height and proportion of vitreous starch in seed endosperm were positively related to seed density. Seedling respiration rate, on a per-seed basis, was associated positively with seed density and size (Lawan *et al.,* 1985). Days from seeding to anthesis decreased from 70 with small, low-density seed, to 62 with large, high-density seed. Seed density has a positive linear relationship with seedling emergence percentages but not with increased yield. Seed size has a major influence on seedling and plant vigour (Gardner and Vanderlip, 1989). Medium-sized millet seed showed higher germination over a wider temperature range than small or large seed. At seven days after planting, seedlings from large and medium-sized high-density seeds were taller than from small or low-density seeds (Mortlock and Vanderlip, 1989).

*Seedling Stage*: In the arid regions, pearl millet is sown with the first onset of monsoon wherein the emerging seedling experiences very high temperatures that affect seed germination, seedling growth and development that ultimately determines the forage and grain yield. The heat stress experienced during the seedling stage affects the photosynthesis, reduces the chlorophyll content, increases the respiration rate thereby causing the death of the seedlings due to excessive transpiration of leaves magnified when accompanied with water stress (Ristic *et al.*, 2007, Cossani and Reynolds, 2012). The earlier studies on photosynthesis indicated that heat stress causes swelling of the thylakoid membrane and malfunction of photosystem II involved in the photosynthetic activity (Ristic *et al.*, 2007; Talukder *et al.,* 2014). Chlorophyll pigment present in the thylakoid membrane is also reduced due to the membrane damaged due to stress (Ristic *et al.,* 2008). The soil temperatures are more as compared to the atmospheric temperatures and pose a challenge for emerging seedlings. In Sahel, the establishment of pearl millet was critical during the first 10 days after sowing that was attributed to high temperatures by Stomph (1990). Similar studies were made by Peacock *et al.* (1993) in the Indian Thar desert where pearl millet is predominantly grown. In the later crop growth stages, transpiration plays a major role in reducing the effect of high temperature by cooling the leaves when not accompanied by drought stress (Stomph, 1990). Though pearl millet is a heat-tolerant crop, the germination rate, establishment, initial growth and photosynthesis rate increases up to a temperature of about 35°C and above that the normal growth is affected (Arya *et al.,* 2014). In almost all the crops, poor stand establishment leads to poor realization of yields. This increases manifold in crops grown under marginal conditions. Singh *et al.* (2003b) studied the seedling heat stress in pearl millet and found that seedling establishment has more relative importance than seedling survival and growth thereby establishing the importance of environmental conditions before emergence. Once the crop is established, any kind of stress occurring at the early stages of plant growth reduces the number of tillers drastically. This triggers the crop to flower early thereby affecting the crop growth and availability of photosynthates that will result in smaller head size, less number of productive tillers which will ultimately cause a reduction in grain yield along with fodder yield (Arya *et al.,* 2010). There is a scope for improving heat tolerance in pearl millet. The genetic variation for seedling survival and thermo tolerance has been reported and it was also found that specific Heat Shock Proteins (HSPs) are involved in the development of thermo tolerance (Soman and Peacock, 1985).

***Vegetative Stage:*** In pearl millet, tillers play a significant role in determining the yield especially in the areas where the plant population is affected due to various abiotic stresses. The rate of tiller production is related to thermal time (Ong, 1984). Effects of temperature on partition of assimilates was studied by Squire (1989). The leaves, stems and panicles of pearl millet were studied for the time taken to gain unit weight (tw), a partition factor (p)—the fraction of new dry matter allocated to the structure during tw. Over a base temperature of 10 °C, as the temperature increased upto 28 to 30 °C, less time was taken by the plant to gain weight in all the plant parts. For stems and panicles, the value of p was, with one exception, little affected by temperature. The dry weight of these plant parts were proportional to tw, and decreased with rise in temperature. For panicles at the temperature of 19°C, p was reduced by 40% as the paicle showed poor seed set. However, for leaves, p increased with rise in temperature, counteracting the effect on tw, thereby the dry weight did not show much change with temperature. The optimum temperature for realizing good grain yield was determined as 22°C though the partitition and mobilization of photosynthates did not show much variation between 22°C and 31°C (Squire, 1989).

Temperature optimum for photosynthesis of intact leaves was reported to be 35-40°C, with about 75% at 45°C (McPherson and Slatyer, 1973). Ashraf and Hafeez (2004) studied thermotolerance in an open pollinated variety of pearl millet, ICMV-94133 at both germination and vegetative stages. The high temperature increased the relative growth rate and net assimilation rates and caused a significant increase in N, P and $K^+$ uptake while the $Ca^{2+}$, $Mg^{2+}$, $Na^+$ and S uptake did not show significant changes. Begg and Burton (1971) reported that with increase in temperatures, leaf umber increased though the crop entered into flowering stage earlier though the drymatter accumulation was higher at the lower temperature. As most of the genotypes reached anthesis sooner at the higher temperature, the diversity was diificult to estimate. At lower temperatures the phenotypic expression was more thereby the variation between genotypes averaged over all photoperiods was greater. The photoperiod influenced the crop growth by increasing the tiller production and the leaf number on the main stem. The photoperiod also governed the effect of temperature on crop growth. At 14h, the influence of temperature was more than at lower photoperiods.

***Flowering to Grain Maturity Stage:*** Pearl millet is grown on a limited scale during summer season in north-western India where irrigation facilities are available. The summer crop occupies an area of more than 500000 ha and is grown from February to June. The climate during this period is hot and dry with air temperatures reaching ≥40°C. Almost all the summer grown area is occupied with hybrids. The high temperatures occurring during flowering stage leads to reduced seed set and poor grain yield in the available hybrids. Evaluation of experimental hybrids led to identification and cultivation of hybrids with good seed set and high grain yield indicating the potential of increasing the yield of summer pearl millet with breeding of heat tolerance during flowering stage (Gupta *et al.,* 2015). High temperatures in the presence of adequate moisture increases the rate of respiration and as reported by Mahalakshmi and Bidinger (1985) affects grain filling, grain yield, grains per unit area and 1000 grain weight when experienced during grain filling stage. In pearl millet, the number of grains is set at second growth stage and grain weight during third growth stage (Maiti and Bidinger, 1981). Panicles per square meter or panicles per plant is the yield component associated with grain yield differences due to temperature (Ong, 1983). Gupta *et al.,* (2015) evaluated six A/B pairs under controlled conditions maintaining a maximum temperature of 43°C and minimum temperature of 22°C. They found that boot-leaf stage of pearl millet plant to be more heat sensitive than panicle-emergence stage indicating that the potential grain yield is set in boot leaf stage itself. The female reproductive structures were more heat sensitive than the pollen. They have compared 23 hybrids ad their parental lines for seed set% ad grain yield at high air temperature (>42°C). Seed set was found to be a dominant trait controlled by non-additive genes. This implies that breeding one of the hybrid parent for heat tolerance would contribute towards heterosis for good seed set and thereby high grain yield during summer season. The parents with good seed set when exposed to high temperatures at flowering can also be used to develop heat tolerant composites. The seed set percentage in pearl millet maintained until the temperature reached 42°C and decreased in curvilinear fashion thereon to 20 percent at 46°C. Similarly, the minimum threshold temperature was found to be 26.4°C while the temperature was observed to be 34.2°C. Similarly, the relationship of percent seed set with vapor pressure deficit (VPD) showed threshold value of 6.2 kPa for maximum VPD, 1.2 kPa for minimum VPD and 3.7 kPa for mean VPD (Gupta *et al.,* 2015).

***Seed Quality:*** Plant productivity is greatly affected when the sensitive growth stages such as reproductive development particularly early seed development are exposed to stresses. The plant tries to complete its life cycle when it is exposed to stresses which is more so when it occurs during the reproductive development. This shorter crop cycle and increased temperatures following fertilization negatively affect grain development (Folsom *et al.,* 2014; Begcy and Walia, 2015; Chen *et al.,* 2016; Begcy and Dresselhaus, 2018). Heat stress during seed development leads to accelerated seed development that results in reduced seed size (Folsom *et al.,* 2014). The process of grain filling is also sensitive to environmental conditions. The high temperature affects the nutrient accumulation in developing grains which has strong effects on final yield and quality (Yang and Zhang, 2006). It also significantly affects seed dormancy, germination, and emergence as well as seedling establishment (Khan, 1976; Finkelstein *et al.,* 2008; Brunel-Muguet *et al.,* 2015). Increased temperatures during grain development has a strong negative effect on seed which decreases its germination potential resulting in reduced seed viability and poor germination (Fahad *et al.,* 2017). Apart from this, the reduction in seed germination and seedling vigor due to HS is associated with reduced thermostability of the plasma membrane as well as membrane fluidity (Saidi *et al.,* 2010; Fahad *et al.,* 2017), which delayed activation of $Ca^{2+}$ signaling, kinases, and heat shock factors (Sangwan *et al.,* 2002; Hofmann, 2009; *Saidi et al.,* 2010). There is a decrease in starch content and a rise in the accumulation of soluble sugars due to abiotic stresses (Krasensky and Jonak, 2012).

## HEAT TOLERANCE MECHANISMS IN CROPS

In order to mitigate the effect of high temperatures, plants have developed both long-term and short-term adaptations (Hong *et al.,* 2003). Survival in hot, dry environments can be achieved in a variety of ways, by combinations of adaptations (Fitter and Hay, 2002). Plant adaptation to heat stress includes escape, avoidance and tolerance mechanisms.

*Escape:* Adjusting life cycle timing is one of the mechanisms used by plants to avoid the abiotic stresses. Such adjustment of life history events (Montesinos-Navarro *et al.,* 2011) is also facilitated by the adjustment of mobilizing the photosynthates to inflorescence (Wolfe and Tonsor, 2014). The pearl millet landraces that are grown still in traditional tracts of arid regions of Rajasthan are early flowering completing their life cycle in 70 to75 days. Such evolutionary adjustments has established the crop serving the food and fodder needs of the marginal farmers. Other than the escape strategy, other adaptive responses adapted by the plants to mitigate the stresses can be described as part of two main strategies, avoidance and tolerance (Sakai and Larcher, 1987).

*Avoidance:* Stress avoidance is a mechanism by which plant makes changes internally to reduce or mitigate the damaging effects of the stresses (Touchette *et al.,* 2009, Puijalon *et al.,* 2011). The mechanisms followed by plants to reduce heat loading or reducing tissue temperature is either through cooling due to transpiration (Earley *et al.,* 2009, Shah *et al.,* 2011), and/or through tolerance, *i.e.* maintaining function at high temperatures (Zhang *et al.,* 2016). Leaves are the main structures of plant that are involved in assimilation and transpiration. However, the transpiration can cool the plant surface only by 4°C as compared to the atmospheric temperature as reported in cotton by Wiegand and Namken (1966). Thus leaf temperature is an indicator of plants' heat avoidance mechanism. Apart from transpiration, temperature can also be controlled by leaf orientation adjustment (Zlatev *et al.,* 2006) wherein plants adjust leaf angle reducing the proportion of leaf area exposed to heat from sunlight (Huey, 2002). The threshold temperature that controls the relative rate of transpiration is species-specific (Mahan *et al.,* 1990). The transpiration-driven heat stress avoidance is a function of both temperature as well as humidity controlled by the stomatal opening. High internal cell temperatures despite avoidance mechanism can damage the cell in many ways. It can cause damage to integrity of cellular structures, affect the protein function and metabolic function by producing ROS as well as cause damage to the membranes (Schöffl *et al.,* 1998) and enzyme denaturation (Ismail and Hall, 1999). The functional disruption is mainly through the reduced photosynthetic rate and activity with increase in temperatures which ultimately determines the plants survival (Senthil-Kumar *et al.,* 2007).

*Tolerance:* When the plants cannot escape or avoid the heat stress, they minimize or repair the damage and this ability to sustain high internal temperatures is heat tolerance (Puijalon *et al.,* 2011). Heat tolerance mechanisms include protection and repair of damaged cell structures, structural proteins, and enzymes (Shah *et al.,* 2011), involve up-regulation of two classes of molecules: heat shock proteins (thereafter, Hsps) (Wang *et al.,* 2004) and plant hormones (Larkindale and Huang, 2005).

***Plant Hormones:***   Plant hormones are known to play an important role in alleviating the damage due to heat stress. Under heat stress, alteration in harmonal homeostasis, stability, content, biosynthesis and compartmentalization are observed (Maestri *et al.,* 2002). Abscisic acid (ABA) accumulated after release of stress and did not show any change during stress period (Larkindale and Huang, 2005). The ABA is involved in the up and down regulation of numerous genes (Xiong *et al.,* 2002). It is also involved in the induction of several HSPs (Pareek *et al.,* 1998) that confirms its involvement in thermo tolerance. Salicylic acid (SA) is a phenolic compound that plays an important role in abiotic stress tolerance as well as disease resistance. It is also involved in the regulation of growth and development of plants under stress conditions through regulation of important plant physiological processes and plant-water relations (Khan *et al.,* 2013). Under heat stress, SA enhanced fresh and dry biomass, concentration of organic and inorganic solutes in rice. Plant functions are controlled by the levels of SA concentrations. Exogenous application has been shown to increase thermo-tolerance. SA was also found to help in overcoming the heat stress affecting spikelet differentiation in rice. SA was reported to induce several genes responsible for encoding chaperone, heat shock proteins (HSPs), antioxidants, and secondary metabolites [sinapyl alcohol dehydrogenase (SAD), cinnamyl alcohol dehydrogenase (CAD), and cytochrome P450 (Jumali *et al.,* 2011)]. SA also played a significant role in reducing the impact of heat stress on photosynthesis by decreasing electrolyte leakage and oxidative stress, alleviated declines in Pn and Rubisco activition state, and improved maximum yield of PSII, Fv/Fm in *Cucumis sativa* seedlings after both heat stress and recovery (Shi *et al.,* 2006). Treatment with 0.5 mM SA can alleviate heat stress in *T. aestivum* through restriction of the stress ethylene formation under heat stress (Khan *et al.* 2013). SA reduces ROS accumulation and affects a great many physiological processes (Khan *et al.,* 2015).

***Secondary Metabolites:***   Different stress conditions gives rise to varied set of metabolites in plants. These variations in metabolite expression are associated with different genotypic and phenotypic traits (Schilmiller *et al.,* 2012). Thus, metabolomics can be used as a selection tool to understand the genetic basis of plant responses to heat stress, understand the genotype based on phenotypic responses and determine the genetic basis of plant responses to heat stress. Metabolites such as tryptophan, serine, threonine, beta-alanine, proline, glutamate, myo-inositol, and urea increased in maize plants under high temperature stress. Certain metabolites showed a negative correlation (threonine, valine, trehalose, glycerol) or a positive correlation (fumarate, succinate, raffinose) with grain yield. The expression of metabolites was same under greenhouse and field conditions (Obata *et al.,* 2015). Condensation of chlorogenoquinone with proteins yielding brown pigments limiting the spread of stress-induced tissue damage in tobacco; formation of phenylamides performing ROS-scavenging ability in tobacco and bean subjected to heat shock is documented. The protective effect of a gaseous secondary metabolite, isoprene, against heat shock was shown, and the ability of isoprene to scavenge singlet oxygen was demonstrated (Edreva *et al.,* 2008). Molmann *et al.* (2015) reported that the concentrations of quercetin and kaempferol in *Medicago sativa* L. plant were higher at the warmer temperature. If heat stress related metabolic biomarkers can be identified for pearl millet, they could be used as targeted and fast diagnostic tools to select germplasm with improved performance under higher temperatures. At present, there are insufficient biomarkers available to adequately and efficiently screen cereal crops for heat stress tolerance or susceptibility.

***Heat Shock Proteins:*** Immediately after exposure to high temperatures and perception of signals, changes occur at the molecular level altering the expression of genes and accumulation of transcripts, thereby leading to the synthesis of stress-related proteins as a stress tolerance strategy (Iba, 2002). Expression of heat shock proteins (HSPs) is known to be an important adaptive strategy in this regard (Feder and Hoffman, 1999). The HSPs, ranging in molecular mass from about 10 to 200 kDa, have chaperonelike functions and are involved in signal transduction during heat stress (Schoffl *et al.,* 1999). The tolerance conferred by HSPs results in improved physiological phenomena such as photosynthesis, assimilate partitioning, water and nutrient use efficiency, and membrane stability (Momcilovic and Ristic, 2007). Such improvements make plant growth and development possible under heat stress. The Hsps have been classified into five groups – Hsp 100, Hsp 90, Hsp 70, Hsp 60 and small Hsps. Members of the Hsp100/ClpB family have shown a significant role in heat tolerance in *Saccharomyces cerevisiae* (Hsp104) (Sanchez and Lindquist, 1990) and *Arabidopsis thaliana* (Hsp101) (Queitsch, 2000a). In pearl millet investigation on heat shock proteins mechanisms were studied (Howarth, 1989) for genotypic and developmental variation during seed germination. It was observed that millet seedlings were competent to synthesize the complete spectrum of HSPs during a two hour's heat shock of 45° C. Further, the investigation indicated that thermo-sensitivity of a genotype and HSPs synthesis was a complex response involving the transcriptional and translational control of RNA in the seed embryo. Reddy *et al.* (2010) isolated a cDNA encoding a cytoplasmic Hsp70 (PgHsc70) from *Pennisetum glaucum* by screening heat-stress cDNA library.

Avoidance and tolerance can have different costs and benefits depending on complex aspects of the growth environment. Thus, different environments may favor tolerance or avoidance in response to the balance of selection acting on the mechanisms involved in each strategy, for irradiance and water availability (Sánchez-Gómez *et al.,* 2006).

## SCREENING METHODOLOGIES

Various abiotic stresses are experienced by crop plants in their life cycle and in arid regions, heat stress is of common occurrence, complex to measure and is subjected to environment influence (Blum, 1988). Due to its quantitative nature, having high genotype × environment, less stability under stress conditions it is cumbersome to produce tolerant genotypes *via* traditional breeding methods (Paulsen, 2002). The high GxE interaction along with difficulties in multi-location testing under heat stress conditions is made more difficult by the low heritability of heat stress tolerance (Lipiec *et al.,* 2013). Also, there is requirement of considerable resources for field screening and suitable environmental conditions for accurate screening and predictable phenotyping performance to distinguish tolerant genotypes. Breeding is usually performed for the measurable sub-traits governing heat stress tolerance rather than tolerance as a single entity. To avoid cumbersome phenotyping method of breeding, the QTLs associated with tolerance are identified, multiple genes or QTLs controlling multiple sub-traits are combined to improve tolerance (Gilliham *et al.,* 2017). The increased solute leakage as a result of decreased cell membrane stability (CMS) is one of the sub-traits used to study drought and heat stress and has been used in many crops to identify tolerant genotypes (Öztürk *et al.,* 2016; Saadalla *et al.,* 1990). CMS can be used to study both drought and heat stress as either of these stresses have similar effects on the plant cell, causing solute leakage due to damage to the plasma membrane which is ultimately detrimental to the plant growth and development (Bajji *et al.,* 2002). The effect of heat stress on the cell membrane's thermal stability was studied under field conditions and *in vitro* by measuring the efflux of electrolytes leaked (Saadalla *et al.,* 1990, Hemantaranjan, 2014). The thermostability of the cell was found to be a under the control of quantitative genes that were moderately heritable (Hemantaranjan, 2014), and showed high genetic correlation with grain yield (Asif and Kamran, 2011). However, it was also shown that a small number of genes controlled a large portion of the variation for this trait (Talukder *et al.,* 2014).

**Field Screening Technique:** Several studies were made to see the effect of temperature on germination. However, the constant temperature that is maintained in the incubators donot correlate with the field studies. This is mainly due the variation in temperature in soil that fluctuates during the day keeping all other factors constant (Singh *et al.,* 2003a). A field screening technique for pearl millet was standardized by Peacock *et al.* (1993) during 1989 and 1990 while screening pearl millet seedlings in arid regions of Rajasthan. The experments were taken up during April and May. Seeds were sown with irrigation and precautions were taken to avoid drought conditions in the field. Environmental measurements were taken for air temperature, soil surface temperature and soil moisture. Seedling emergence was recorded when the first seedlings were seen in the field and data were recorded daily at 1700 local time until no further emergence occurred. The number of live seedlings was counted daily and dead seedlings were marked with wooden sticks to check against loss of due to damage by birds or rodents. The ratio of seedling survived to the total number of seedlings emerged gave TI. Good genetic variation was obtained with respect to TI and the genotypes were grouped as tolerant and susceptible to heat. Using this technique a large number of genotypes can be screened at the same time. Later on Singh (1993) used the same technique and proposed SSTI which takes both, pre and post-emergence mortalities into account. The difference in germination under stress and normal conditions gives preemergence mortality which may confound the TI. In pearl millet, the genes govering grain yield under stress differed from those governing grain yield under optimal growth conditions which emphasizes that there is a need to identify representative target environments ad practice selection in those environments to improve the yield of the pearl millet genotypes under heat stress (Patill and Jadeja, 2009; Yadav *et al.,* 2012). As pre-emergence mortality is considered, SSTI was found to be effective over STI for testing of varieties in the advanced trials (Yadav *et al.,* 2013).

**Laboratory-cum-Field Screening Technique**: Though the laboratory responses are not highly correlated to field responses, genotypic response to temperature is difficult to study in the field due to lack of uniformity of the soil with respect to moisture status and thereby temperature as well as the crust development which makes it even more complex (Soman *et al.,* 1984). Laboratory techniques representing near field conditions have also been developed and used for sorghum and pearl millet. One such technique was developed by Soman and Peacock (1985). Clay pots were filled with sieved top soil and placed in steel water tank so that top 7 cm of soil are above water level. Also, 2 cm gap existed from the rim of the pot to the surface of soil. The surface of the water was covered with floating white plastic balls to reduce evaporation. The soil in pots was heated with infra red lamps (240-250 V, 250 W, Phillips, Type IRR) fitted to frame above the tank in two rows. The height of the frame of infra red lamps was adjusted to vary the temperature of the soil surface. The study was conducted at 35, 40, 45 and 50° C measured at 2 cm below the soil surface. Water level was maintained by monitoring every day ensuring that the wet soil column (25 cm long) in the pots and soil water content was measured gravimetrically. Plants were subjected to heat stress while avoiding moisture stress. The experiment showed that a temperature of 45° C could distinct between tolerant and susceptible genotypes while very few lines emerged at 50° C (Yadav *et al.,* 2006).

**Laboratory Techniques:** Laboratory techniques have been developed and widely used by various scientists in many crops. Membrane thermo stability is one such trait that is widely used for measuring the high temperature tolerance through measurement of electrolyte leakage of leaves subjected to heat stress (Shanahan *et al.,* 1990). The laboratory methods are classified as the electrical conductivity methods (Nagarajan and Panda, 1980), Chlorophyll florescence method (Smillie and Gibbons, 1981) and embryo protein synthesis methods (Ougham *et al.,* 1988). Three screening techniques *viz.,* (1) seed germination in polyethylene glycol-600 (PEG) at -0.6 MPa osmotic pressure, (2) seedlings subjected to PEG-600 at -0.06 MPa osmotic pressure in hydroponic solution for 14 days and (3) a cellular MTS test were compared by Bouslama and Schapaugh (1984). Cultivars having both heat and drought tolerance were identified and significant correlation observed between hydroponic screening and heat tolerance tests. Heat stress affected the growth and yield of cultivars by affecting the physiological aspects like MTS and chlorophyll fluorescence, regulation of compatible solutes (Takeda *et al.,* 1999). There was also an accumulation of free proline in the leaves in response to high temperature.

Damage to membrane assayed by the MTS is an old and reliable screening trait for heat stress tolerance that is used in several crops (Sullivan, 1972). Although MTS is positively associated with yield performance in pearl millet under heat and drought stressed conditions (Yadav *et al.,* 2009), the identified heat tolerance indices STI and SSTI did not show stable pooled correlation with developmental traits except germination and emergence rate. After a series of studies, it was suggested that MTS may be used for screening large number of genotypes while field-based indices STI and SSTI may be used for evaluation of hybrids and varieties before they are released (Yadav *et al.,* 2014).

## BREEDING FOR HEAT TOLERANCE

Pearl millet is a cross-pollinated crop with >75 per cent out-crossing. The genetic improvement is through recurrent selection procedures. Arid regions are mostly grown with landraces or OPVs while hybrids are preferred in productive regions which has been feasible with the availability of CMS system (Sanjana, 2015).

While breeding for heat tolerance in crops, the initial search for tolerant sources should involve both advanced breeding lines and landraces of the species. Significant genetic diversity was found among the 105 landraces of pearl millet (*Pennisetum glaucum*) collected from north-western India by Yadav *et al.* (2004).

The landraces are grown by farmers over a period of time and which have acquired the necessary adaptation genes and thus could be potential source for heat tolerance. Such sources need to be first screened through established screening methods. They have to be fixed through selfing and sibbing and later on introgressed into breeding material with desirable genetic background suitable for the region. Early flowering, tillering, high biomass, efficient partitioning of photosynthates was crucial in determining grain productivity under arid zone. The manifestation of heterosis in the landrace-based topcross hybrids varied for different traits and significant heterosis for biomass, grain yield and stover yield was observed in specific male-sterile seed parent × landrace-based pollinator combinations (Yadav and Bidinger, 2008). Till recent, escape mechanism is widely used where selection for early flowering and maturity has enabled to escape heat and drought stress in pearl millet. There is also a choice of utilizing crop wild relatives to breed for tolerance, but their incorporation into cultivated background needs a perfect prebreeding/backcross program and also good sources for tolerance in the form of landraces are available.

Though it is seen in several studies that the direct selection for yield improvement under stressed conditions is the most effective way for realising improved cultivars with heat stress tolerance, low heritability of the traits governing heat tolerance and a complex network of major and minor QTLs governing them hinder the progress in varying climatic conditions across the years (Leung, 2008; Manavalan *et al.,* 2009). Initial screening for stress tolerance also lead to losing valuable yield genes. There is a need for good understanding of genetic inheritance of HT, and development of validated QTLs/cloned gene(s) for HT for achieving rapid gains (Cossani and Reynolds, 2012).

Heat stress tolerance at seedling stage in pearl millet was studied extensively by Yadav *et al.* (2011). They have listed the morphological traits governing seedling heat tolerance such as seedling heat tolerance index, seed to seedling heat tolerance index, emergence rate, leaves/seedling, seedling height, seedling fresh and dry weights. These traits were found to be under the control of non-additive gene action that can be exploited through development of hybrids (Yadav *et al.,* 2011). The estimates of additive genetic variance and narrow sense heritability for STI and SSTI were high (Yadav *et al.,* 2011). Patil and Jadeja (2009) repoted that alleles controlling the grain yield in stress and non-stress environment were partially different, therefore, need to select in target environment to improve the performance of the genotypes. Hybrids were found to perform better than the parents with respect to the heat tolerance and yield traits (Joshi *et al.,* 1997). The genotypes CVJ 2-5-3-1-3 and 77/371// BSECP CP-1 were identified as the best general combiners for heat tolerance indices. The heat tolerance at seedling stage had no significant impact on its later growth and development (Yadav *et al.,*

2006). the The importance of additive gene effects in acquired thermal tolerance was also stressed by Ibrahim and Quick (2001). The range for STI in the hybrids was from 53 to 82 while SSTI values from 46 to 73. Both these heat tolerance indices did not show significant correlation with the maturity traits suggesting the recovering ability of pearl millet though the effect due to poor stand cannot be ignored (Yadav *et al.,* 2012). The important physiological traits that need to be considered include those related to canopy structure, delayed senescence, photosynthesis efficiency, less respiration rates, reproductive traits, and harvest index (Cossani and Reynolds, 2012; Gupta *et al.,* 2012). However, not much progress has been made to deploy these traits in breeding programs of pearl millet.

Large genetic variation for tolerance to heat at reproductive stage has been observed and heat tolerant lines have been identified such as ICMB 92777, ICMB 05666, ICMB 00333, ICMB 01888, ICMB 02333 and ICMB 03555 among the maintainer lines and ICMV 82132, MC 94, ICTP 8202 and MC-bulk among the populations (Yadav *et al.,* 2012). Based on 3 to 4 year field screening (2009–2012), five hybrid seed parents (ICMB 92777, ICMB 05666, ICMB 00333, ICMB 02333 and ICMB 03555) and a germplasm accession IP 19877 with 61 to 69% seed set as compared to 71% seed set in a heat tolerant commercial hybrid 9444 (used as a control) was identified (Gupta *et al.,* 2015). Hybrids outperformed the others in terms of seed set and grain yield for which CZH 233, CZP 9603, CZI 2011/5 and CZMS 21A were the best performing genotypes (Aravind *et al.,* 2017).

Molecular markers are preferred for transferring quantitative traits into desirable genotypes that are difficult to screen in field conditions in a quicker and precise manner (Shirasawa *et al.,* 2013). Once the markers associated with QTLs have been identified, the candidate QTLs can further be introgressed into elite lines through marker-assisted selection (MAS) strategies. Traits governing heat stress or in a matter of fact any stress are mostly governed by small effect QTLs or several epistatic QTLs which makes it difficult to introgress (Bita and Gerats, 2013). There is a need to pyramid several QTLs in the same genetic background using large populations through marker-assisted recurrent selection (MARS) or genomic selection (GS) to achieve desirable levels of resistance. MAS programs for complex traits such as heat tolerance are not effective mainly due to the genotype × environment and gene-gene (*i.e.,* epistasis) interactions, which frequently result in a low breeding efficiency (Collins *et al.,* 2008). Using segregating populations derived from two crosses H 77/833-2 and PRLT 2/89-33, and ICMB 841 and 863B, a major QTL for terminal drought tolerance in pearl millet has been identified and mapped on linkage group 2 (LG 2) (Bidinger *et al.,* 2007). This QTL on LG 2 is being targeted for increasing grain yield and grain stability for terminal drought tolerance in targeted environments (Yadav et al., 2011). This DT-QTL has also been found to confer high leaf abscisic acid (ABA) and limiting transpiration rates at high vapor pressure deficits supporting the hypothesis of water saving mechanisms under well-watered conditions in drought tolerant pearl millet lines (Kholová *et al.,* 2010).

In contrast to MAS strategies where markers linked to target QTLs are identified and used, the GS method predicts breeding values using data derived from a vast number of molecular markers with a high coverage of the genome. Its novelty is that it uses all marker data as predictors of performance and subsequently delivers more accurate predictions (Bita and Gerats, 2013). Simulation studies indicated that GS may increase the correlations between predicted and true breeding value over several generations, without the need to re-phenotype. Thus, GS may result in lower analysis costs and increased rates of genetic gain (Habier *et al.,* 2009; Heffner *et al.,* 2009). Precision phenotyping protocols are important for initiating any breeding process. The standardized protocols and trials is a prerequisite for enabling the assessment of the complex genetic networks associated with QTLs.

There has been only one report on transcriptome analysis of pearl millet under abiotic stress conditions (Mishra *et al.,* 2007). A total of 2,494 differentially regulated transcripts in response to drought, salinity and cold stress were identified and the study indicated the existence of a complex gene regulatory network that differentially modulates gene expression under various stresses. Despite its large genome size, pearl millet genome has been sequenced and comprehensive transcriptome has been developed by combining together the transcriptome data of three independent research works namely Rajaram *et al.* (2013), Zeng *et al.* (2011) and Yves *et al.,* (unpublished; http://www.ceg. icrisat.org/ transcriptome.html).

## CONCLUSIONS

Due to its inherent stress tolerant nature, pearl millet is grown in arid regions where heat stress at germination and seedling stage hinders the crop establishment. Heat tolerance at anthesis stage is required in areas where pearl millet is grown during summer season. Not much work has been done in breeding for heat tolerance. Tolerant sources and traits governing resistance have been identified. Screening methodology has been standardized. Most of the traits are governed by dominance gene action. There is a need for understanding the underlying component traits and their inheritance, identify the genes/QTLs and pyramiding the genes into a single genotype.

## CONSENT FOR PUBLICATION

Not applicable.

## CONFLICT OF INTEREST

The authors confirm that this chapter content has no conflict of interest.

## ACKNOWLEDGEMENTS

Declared none.

## REFERENCES

Aravind, J., Manga, V.K., Bhatt, R.K., Pathak, R. (2017). Differential response of pearl millet genotypes to high temperature stress at flowering. *J. Environ. Biol., 38*(5), 791-797.
[http://dx.doi.org/10.22438/jeb/38/5/MRN-432]

Arya, R.K., Yadav, H.P., Yadav, A.K., Singh, M.K. (2010). Effect of environment on yield and its contributing traits in pearl millet. *Forage Res., 36*, 176-180.

Arya, R.K., Singh, M.K., Yadav, A.K. (2014). Advances in pearl millet to mitigate adverse environment conditions emerged due to global warming. *Forage Res., 40*(2), 57-70.

Ashraf, M., Hafeez, M. (2004). Thermotolerance of pearl millet and maize at early growth stages: growth and nutrient relations. *Biol. Plant., 48*(1), 81-86.
[http://dx.doi.org/10.1023/B:BIOP.0000024279.44013.61]

Asif, M., Kamran, A. (2011). Plant breeding for water-limited environments. *Crop Sci., 51*, 2911.
[http://dx.doi.org/10.2135/cropsci2011.12.0004br]

Bajji, M., Kinet, J.M., Lutts, S. (2002). The use of the electrolyte leakage method for assessing cell membrane stability as a water stress tolerance test in durum wheat. *Plant Growth Regul., 36*, 61-70.
[http://dx.doi.org/10.1023/A:1014732714549]

Begcy, K., Dresselhaus, T. (2018). Epigenetic responses to abiotic stresses during reproductive development in cereals. *Plant Reprod., 31*(4), 343-355.
[http://dx.doi.org/10.1007/s00497-018-0343-4] [PMID: 29943158]

Begcy, K., Walia, H. (2015). Drought stress delays endosperm development and misregulates genes associated with cytoskeleton organization and grain quality proteins in developing wheat seeds. *Plant Sci., 240*, 109-119.
[http://dx.doi.org/10.1016/j.plantsci.2015.08.024] [PMID: 26475192]

Begg, J.E., Burton, G.W. (1971). Comperative study of five genotypes of pearl millet under a range of photoperiods and temperatures. *Crop Sci., 11*, 803-805.
[http://dx.doi.org/10.2135/cropsci1971.0011183X001100060009x]

Bidinger, F.R., Nepolean, T., Hash, C.T., Yadav, R.S., Howarth, C.J. (2007). Identification of QTLs for grain yield of pearl millet [ *Pennisetum glaucum* (L.) R. Br.] in environments with variable moisture during grain filling. *Crop Sci., 47*, 969-980.
[http://dx.doi.org/10.2135/cropsci2006.07.0465]

Bita, C.E., Gerats, T. (2013). Plant tolerance to high temperature in a changing environment: scientific fundamentals and production of heat stress-tolerant crops. *Front. Plant Sci., 4*(273), 273.
[http://dx.doi.org/10.3389/fpls.2013.00273] [PMID: 23914193]

Blum, A. (1988). *Plant Breeding for Stress Environments.* Boca Raton, Florida: CRC Press Inc. (p. 223).

Bouslama, M., Schapaugh, W.T. (1984). Stress tolerance in soybeans. I. Evaluation of three screening techniques for heat and drought tolerance. *Crop Sci., 24*, 933-937.
[http://dx.doi.org/10.2135/cropsci1984.0011183X002400050026x]

Brunel-Muguet, S., D'Hooghe, P., Bataillé, M.P., Larré, C., Kim, T.H., Trouverie, J., Avice, J.C., Etienne, P., Dürr, C. (2015). Heat stress during seed filling interferes with sulfur restriction on grain composition and seed germination in oilseed rape (*Brassica napus* L.). *Front. Plant Sci., 6*, 213.
[http://dx.doi.org/10.3389/fpls.2015.00213] [PMID: 25914702]

Chen, C., Begcy, K., Liu, K., Folsom, J.J., Wang, Z., Zhang, C., Walia, H. (2016). Heat stress yields a unique MADS box transcription factor in determining seed size and thermal sensitivity. *Plant Physiol., 171*(1), 606-622.
[http://dx.doi.org/10.1104/pp.15.01992] [PMID: 26936896]

Collins, N.C., Tardieu, F., Tuberosa, R. (2008). Quantitative trait loci and crop performance under abiotic stress: where do we stand? *Plant Physiol., 147*(2), 469-486.
[http://dx.doi.org/10.1104/pp.108.118117] [PMID: 18524878]

Cooper, P., Rao, K.P.C., Singh, P., Dimes, J., Traore, P.S., Rao, K., Dixit, P., Twomlow, S. (2009). Farming with current and future climate risk: advancing a 'hypothesis of hope' for rain-fed agriculture in the semi-arid tropics. *J. SAT Agric. Res., 7*, 1-19.

Cossani, C.M., Reynolds, M.P. (2012). Physiological traits for improving heat tolerance in wheat. *Plant Physiol., 160*(4), 1710-1718.
[http://dx.doi.org/10.1104/pp.112.207753] [PMID: 23054564]

Das, A., Rushton, P.J., Rohila, J.S. (2017). Metabolic profiling of soybeans (*Glycine max* L.) reveals the importance of sugar and nitrogen metabolism under drought and heat stress. *Plants (Basel), 6*(2), 21.
[http://dx.doi.org/10.3390/plants6020021] [PMID: 28587097]

Directorate of Millets Development. (2019). http://millets.dacfw.nic.in/India_Apy.html

Driedonks, N., Rieu, I., Vriezen, W.H. (2016). Breeding for plant heat tolerance at vegetative and reproductive stages. *Plant Reprod., 29*(1-2), 67-79.
[http://dx.doi.org/10.1007/s00497-016-0275-9] [PMID: 26874710]

Earley, E.J., Ingland, B., Winkler, J., Tonsor, S.J. (2009). Inflorescences contribute more than rosettes to lifetime carbon gain in *Arabidopsis thaliana* (Brassicaceae). *Am. J. Bot., 96*(4), 786-792.
[http://dx.doi.org/10.3732/ajb.0800149] [PMID: 21628233]

Edreva, A., Velikova, V., Tsonev, T., Dagnon, S., Gurel, A., Aktas, L., Gesheva, E. (2008). Stress-protective role of secondary metabolites: diversity of functions and mechanisms. *Gen. Appl. Plant Physiol., 34*(1-2), 67-78.

Fahad, S., Bajwa, A.A., Nazir, U., Anjum, S.A., Farooq, A., Zohaib, A., Sadia, S., Nasim, W., Adkins, S., Saud, S., Ihsan, M.Z., Alharby, H., Wu, C., Wang, D., Huang, J. (2017). Crop production under drought and heat stress: plant responses and management options. *Front. Plant Sci., 8*, 1147.
[http://dx.doi.org/10.3389/fpls.2017.01147] [PMID: 28706531]

Feder, M.E., Hofmann, G.E. (1999). Heat-shock proteins, molecular chaperones, and the stress response: evolutionary and ecological physiology. *Annu. Rev. Physiol., 61*, 243-282.
[http://dx.doi.org/10.1146/annurev.physiol.61.1.243] [PMID: 10099689]

Finkelstein, R., Reeves, W., Ariizumi, T., Steber, C. (2008). Molecular aspects of seed dormancy. *Annu. Rev. Plant Biol., 59*, 387-415.
[http://dx.doi.org/10.1146/annurev.arplant.59.032607.092740] [PMID: 18257711]

Fitter, A.H., Hay, R.K.M. (2002). *Environmental Physiology of Plants.* (3rd ed.). London, UK: Academic Press.

Folsom, J.J., Begcy, K., Hao, X., Wang, D., Walia, H. (2014). Rice fertilization-Independent Endosperm1 regulates seed size under heat stress by controlling early endosperm development. *Plant Physiol., 165*(1),

238-248.
[http://dx.doi.org/10.1104/pp.113.232413] [PMID: 24590858]

Gardner, G.C., Vanderlip, R.L. (1989). Seed size and density effects on field performance of pearl millet. *Trans. Kans. Acad. Sci., 92*(1-2), 49-59.
[http://dx.doi.org/10.2307/3628189]

Gilliham, M., Able, J.A., Roy, S.J. (2017). Translating knowledge about abiotic stress tolerance to breeding programmes. *Plant J., 90*(5), 898-917.
[http://dx.doi.org/10.1111/tpj.13456] [PMID: 27987327]

Gupta, P.K., Balyan, H.S., Gahlaut, V., Kulwal, P.L. (2012). Phenotyping, genetic dissection, and breeding for drought and heat tolerance in common wheat: status and prospects. *Plant Breed. Rev., 36*, 85-168.
[http://dx.doi.org/10.1002/9781118358566.ch2]

Gupta, S.K., Rai, K.N. (2015). Seed set variability under high temperatures during flowering period in pearl millet (*Pennisetum glaucum* L. (R.) Br.). *Field Crops Res., 171*, 41-53.
[http://dx.doi.org/10.1016/j.fcr.2014.11.005]

Gupta, S.K., Rai, K.N., Singh, P., Ameta, V.L., Gupta, S.K., Jayalekha, A.K., Mahala, R.S., Pareek, S., Swami, M.L., Verma, Y.S. (2015). Seed set variability under high temperatures during flowering period in pearl millet (Pennisetum glaucum L. (R.) Br.). *Field Crops Res., 171*, 41-53.

Habier, D., Fernando, R.L., Dekkers, J.C.M. (2009). Genomic selection using low-density marker panels. *Genetics, 182*(1), 343-353.
[http://dx.doi.org/10.1534/genetics.108.100289] [PMID: 19299339]

Hall, A.E. (1992). Breeding for heat tolerance. *Plant Breed. Rev., 10*, 129-168.

Hasanuzzaman, M., Nahar, K., Alam, M.M., Roychowdhury, R., Fujita, M. (2013). Physiological, biochemical, and molecular mechanisms of heat stress tolerance in plants. *Int. J. Mol. Sci., 14*(5), 9643-9684.
[http://dx.doi.org/10.3390/ijms14059643] [PMID: 23644891]

Heffner, E.L., Sorrells, M.E., Jannink, J.L. (2009). Genomic selection for crop improvement. *Crop Sci., 49*, 1-12.
[http://dx.doi.org/10.2135/cropsci2008.08.0512]

Hemantaranjan, A. (2014). Heat Stress Responses and Thermotolerance. *Adv. Plants Agric. Res., 1*, 1-10.
[http://dx.doi.org/10.15406/apar.2014.01.00012]

Hofmann, N.R. (2009). The plasma membrane as first responder to heat stress. *Plant Cell, 21*(9), 2544.
[http://dx.doi.org/10.1105/tpc.109.210912] [PMID: 19773384]

Hong, S.W., Lee, U., Vierling, E. (2003). Arabidopsis hot mutants define multiple functions required for acclimation to high temperatures. *Plant Physiol., 132*(2), 757-767.
[http://dx.doi.org/10.1104/pp.102.017145] [PMID: 12805605]

Howarth, C. (1989). Heat shock proteins in *Sorghum bicolor* and *Pennisetum americanum* I. genotypic and developmental variation during seed germination. *Plant Cell Environ., 12*(5), 471-477.
[http://dx.doi.org/10.1111/j.1365-3040.1989.tb02119.x]

Howarth, C.J., Pollock, C.J., Peacock, J.M. (1997). Development of laboratory-based methods for assessing seedling thermo tolerance in pearl millet. *New Phytol., 137*, 129-139.
[http://dx.doi.org/10.1046/j.1469-8137.1997.00827.x]

Huey, R.B., Carlson, M., Crozier, L., Frazier, M., Hamilton, H., Harley, C., Hoang, A., Kingsolver, J.G. (2002). Plants *versus* animals: do they deal with stress in different ways? *Integr. Comp. Biol., 42*(3), 415-423.
[http://dx.doi.org/10.1093/icb/42.3.415] [PMID: 21708736]

Iba, K. (2002). Acclimative response to temperature stress in higher plants: approaches of gene engineering for temperature tolerance. *Annu. Rev. Plant Biol., 53*, 225-245.
[http://dx.doi.org/10.1146/annurev.arplant.53.100201.160729] [PMID: 12221974]

Ibrahim, A.M.H., Quick, J.S. (2001). Genetic control of high temperature tolerance in wheat as measured by membrane thermal stability. *Crop Sci., 41*, 1405-1407.
[http://dx.doi.org/10.2135/cropsci2001.4151405x]

Ismail, A.M., Hall, A.E. (1999). Reproductive-stage heat tolerance, leaf membrane thermostability and plant morphology in cowpea. *Crop Sci., 39*, 1762-1768.
[http://dx.doi.org/10.2135/cropsci1999.3961762x]

Jumali, S.S., Said, I.M., Ismail, I., Zainal, Z. (2011). Genes induced by high concentration of salicylic acid in *Mitragyna speciosa*. *Aust. J. Crop Sci., 5*, 296.

Joshi, A.K., Pandya, J.N., Mathukia, R.K., Pethani, K.V., Dave, H.R. (1997). Seed germination in pearl millet hybrids and parents under extreme temperature conditions. *GAU Res. J., 23*, 77-83.

Khan, M.I.R., Iqbal, N., Masood, A., Per, T.S., Khan, N.A. (2013). Salicylic acid alleviates adverse effects of heat stress on photosynthesis through changes in proline production and ethylene formation. *Plant Signal. Behav., 8*(11), e26374
[http://dx.doi.org/10.4161/psb.26374] [PMID: 24022274]

Khan, M.I.R., Fatma, M., Per, T.S., Anjum, N.A., Khan, N.A. (2015). Salicylic acid-induced abiotic stress tolerance and underlying mechanisms in plants. *Front. Plant Sci., 6*, 462.
[http://dx.doi.org/10.3389/fpls.2015.00462] [PMID: 26175738]

Khan, R.A. (1976). Effect of high-temperature stress on the growth and seed characteristics of barley and cotton. *Basic Life Sci., 8*, 319-324.
[PMID: 1032107]

Kholová, J., Hash, C.T., Kumar, P.L., Yadav, R.S., Kocová, M., Vadez, V. (2010). Terminal drought-tolerant pearl millet [*Pennisetum glaucum* (L.) R. Br.] have high leaf ABA and limit transpiration at high vapour pressure deficit. *J. Exp. Bot., 61*(5), 1431-1440.
[http://dx.doi.org/10.1093/jxb/erq013] [PMID: 20142425]

Krasensky, J., Jonak, C. (2012). Drought, salt, and temperature stress-induced metabolic rearrangements and regulatory networks. *J. Exp. Bot., 63*(4), 1593-1608.
[http://dx.doi.org/10.1093/jxb/err460] [PMID: 22291134]

Larkindale, J., Huang, B. (2005). Effects of abscisic acid, salicylic acid, ethylene and hydrogen peroxide in thermotolerance and recovery for creeping bentgrass. *Plant Growth Regul., 47*, 17-28.
[http://dx.doi.org/10.1007/s10725-005-1536-z]

Lawan, M., Barnett, F.L., Khaleeq, B., Vanderlip, R.L. (1985). Seed density and seed size of pearl millet as related to field emergence and several seed and seedling traits. *Agron. J., 77*(4), 567-571.
[http://dx.doi.org/10.2134/agronj1985.00021962007700040015x]

Leung, H. (2008). Stressed genomics-bringing relief to rice fields. *Curr. Opin. Plant Biol., 11*(2), 201-208.
[http://dx.doi.org/10.1016/j.pbi.2007.12.005] [PMID: 18294900]

Lipiec, J., Doussan, C., Nosalewicz, A., Kondracka, K. (2013). Effect of drought and heat stresses on plant growth and yield: A review. *Int. Agrophys., 27*, 463-477.
[http://dx.doi.org/10.2478/intag-2013-0017]

Lobell, D.B., Asner, G.P. (2003). Climate and management contributions to recent trends in U.S. agricultural yields. *Science, 299*(5609), 1032.
[http://dx.doi.org/10.1126/science.1077838] [PMID: 12586935]

Maestri, E., Klueva, N., Perrotta, C., Gulli, M., Nguyen, H.T., Marmiroli, N. (2002). Molecular genetics of heat tolerance and heat shock proteins in cereals. *Plant Mol. Biol., 48*(5-6), 667-681.
[http://dx.doi.org/10.1023/A:1014826730024] [PMID: 11999842]

Mahalakshmi, V., Bidinger, F.R. (1985). Flowering response of pearl millet to water stress during panicle development. *Ann. Appl. Biol., 106*, 571-578.
[http://dx.doi.org/10.1111/j.1744-7348.1985.tb03148.x]

Mahan, J.R., Burke, J.J., Orzech, K.A. (1990). Thermal Dependence of the Apparent K(m) of Glutathione Reductases from Three Plant Species. *Plant Physiol., 93*(2), 822-824.
[http://dx.doi.org/10.1104/pp.93.2.822] [PMID: 16667543]

Maiti, R.K., Bidinger, F.R. (1981). Growth and development of pearl millet plant. *ICRISAT Res. Bull.* Hyderabad, India. (Vol. 6)

Manavalan, L.P., Guttikonda, S.K., Tran, L.S., Nguyen, H.T. (2009). Physiological and molecular approaches to improve drought resistance in soybean. *Plant Cell Physiol., 50*(7), 1260-1276.
[http://dx.doi.org/10.1093/pcp/pcp082] [PMID: 19546148]

McPherson, H.G., Slatyer, R.O. (1973). Mechanisms regulating photosynthesis in *Pennisetum typhoides*. *Aust. J. Biol. Sci., 26*, 329-339.
[http://dx.doi.org/10.1071/BI9730329]

Mishra, R.N., Reddy, P.S., Nair, S., Markandeya, G., Reddy, A.R., Sopory, S.K., Reddy, M.K. (2007). Isolation and characterization of expressed sequence tags (ESTs) from subtracted cDNA libraries of *Pennisetum glaucum* seedlings. *Plant Mol. Biol., 64*(6), 713-732.
[http://dx.doi.org/10.1007/s11103-007-9193-4] [PMID: 17558562]

Mølmann, J.A.B., Steindal, A.L.H., Bengtsson, G.B., Seljåsen, R., Lea, P., Skaret, J., Johansen, T.J. (2015). Effects of temperature and photoperiod on sensory quality and contents of glucosinolates, flavonols and vitamin C in broccoli florets. *Food Chem., 172*, 47-55.
[http://dx.doi.org/10.1016/j.foodchem.2014.09.015] [PMID: 25442522]

Momcilovic, I., Ristic, Z. (2007). Expression of chloroplast protein synthesis elongation factor, EF-Tu, in two lines of maize with contrasting tolerance to heat stress during early stages of plant development. *J. Plant Physiol., 164*(1), 90-99.
[http://dx.doi.org/10.1016/j.jplph.2006.01.010] [PMID: 16542752]

Montesinos-Navarro, A., Wig, J., Pico, F.X., Tonsor, S.J. (2011). Arabidopsis thaliana populations show clinal variation in a climatic gradient associated with altitude. *New Phytol., 189*(1), 282-294.
[http://dx.doi.org/10.1111/j.1469-8137.2010.03479.x] [PMID: 20880224]

Mortlock, M.Y., Vanderlip, R.L. (1989). Germination and establishment of pearl millet and sorghum of different seed qualities under controlled high-temperature environments. *Field Crops Res., 22*(3), 195-209.
[http://dx.doi.org/10.1016/0378-4290(89)90092-0]

Nagarajan, S., Panda, B.C. (1980). On evaluation of high temperature tolerance of crop plants by electrical conductivity methods. *Indian J. Exp. Biol., 18*(11), 1174-1178.

Obata, T., Witt, S., Lisec, J., Palacios-Rojas, N., Florez-Sarasa, I., Yousfi, S., Araus, J.L., Cairns, J.E., Fernie, A.R. (2015). Metabolite profiles of maize leaves in drought, heat and combined stress field trials reveal the relationship between metabolism and grain yield. *Plant Physiol., 169*(4), 2665-2683.
[http://dx.doi.org/10.1104/pp.15.01164] [PMID: 26424159]

Ong, C.K. (1984). Responses to temperature in a stand of pearl millet (*Pennisetum typhoides* S & H). *J. Exp. Bot., 35*, 83-90.
[http://dx.doi.org/10.1093/jxb/35.1.83]

Ong, C.K. (1983). Response to temperature in a stand of pearl millet (*Pennisetum typhoides* S. and H.). I. Vegetative development. *J. Exp. Bot., 34*, 322-336.
[http://dx.doi.org/10.1093/jxb/34.3.322]

Ortiz, R., Sayre, K.D., Govaerts, B., Gupta, R., Subbarao, G.V., Ban, T., Hodson, D., Dixon, J.M., Ortiz-Monasterio, J.I., Reynolds, M. (2008). Climate change: Can wheat beat the heat? *Agric. Ecosyst. Environ., 126*, 46-58.
[http://dx.doi.org/10.1016/j.agee.2008.01.019]

Ougham, H.J., Peacock, J.M., Staddart, T.L., Soman, P. (1988). High temperature effects on seedling emergence and embryo protein synthesis of Sorghum. *Crop Sci., 8*, 251-254.

[http://dx.doi.org/10.2135/cropsci1988.0011183X002800020014x]

Öztürk, A. (2016). Evaluation of bread wheat genotypes for early drought resistance *via* germination under osmotic stress, cell membrane damage, and paraquat tolerance. *Turk. J. Agric., 40*, 146-159.
[http://dx.doi.org/10.3906/tar-1501-136]

Pareek, A., Singla, S., Grover, A. (1998). Proteins alterations associated with salinity, desiccation, high and low temperature stresses and abscisic acid application in seedlings of pusa 169, a high-yielding rice (*Oryza sativa* L.) cultivar. *Curr. Sci., 75*, 1023-1035.

Patil, H.E., Jadeja, G.C. (2009). Comparative Studies Character Association of Yield and Yield Component Under Irrigated and Terminal Drought Condition in Pearl Millet (*Pennisetum glaucum* L). *Indian Journal of Dryland Agricultural Research and Development, 24*, 53-58.

Paulsen, G. (2002). Application of physiology in wheat breeding. *Crop Sci., 42*, 2228.
[http://dx.doi.org/10.2135/cropsci2002.2228]

Peacock, J.M.P., Soman, R., Jayachandran, A.V., Rani, C.J. (1993). Effect of high soil surface temperature on seedling survival in pearl millet. *Exp. Agric., 29*, 215-225.
[http://dx.doi.org/10.1017/S0014479700020664]

Pearson, C.J. (1975). Thermal adaptation of *Pennisetum* seedling development. *Aust. J. Plant Physiol., 2*, 413-424.

Puijalon, S., Bouma, T.J., Douady, C.J., van Groenendael, J., Anten, N.P., Martel, E., Bornette, G. (2011). Plant resistance to mechanical stress: evidence of an avoidance-tolerance trade-off. *New Phytol., 191*(4), 1141-1149.
[http://dx.doi.org/10.1111/j.1469-8137.2011.03763.x] [PMID: 21585390]

Queitsch, C., Hong, S.W., Vierling, E., Lindquist, S. (2000). Heat shock protein 101 plays a crucial role in thermotolerance in Arabidopsis. *Plant Cell, 12*(4), 479-492.
[http://dx.doi.org/10.1105/tpc.12.4.479] [PMID: 10760238]

Rajaram, V., Nepolean, T., Senthilvel, S., Varshney, R.K., Vadez, V., Srivastava, R.K., Shah, T.M., Supriya, A., Kumar, S., Ramana Kumari, B., Bhanuprakash, A., Narasu, M.L., Riera-Lizarazu, O., Hash, C.T. (2013). Pearl millet [Pennisetum glaucum (L.) R. Br.] consensus linkage map constructed using four RIL mapping populations and newly developed EST-SSRs. *BMC Genomics, 14*, 159.
[http://dx.doi.org/10.1186/1471-2164-14-159] [PMID: 23497368]

Reddy, P.S., Mallikarjuna, G., Kaul, T., Chakradhar, T., Mishra, R.N., Sopory, S.K., Reddy, M.K. (2010). Molecular cloning and characterization of gene encoding for cytoplasmic Hsc70 from *Pennisetum glaucum* may play a protective role against abiotic stresses. *Mol. Genet. Genomics, 283*(3), 243-254.
[http://dx.doi.org/10.1007/s00438-010-0518-7] [PMID: 20127116]

Ristic, Z., Bukovnik, U., Momcilović, I., Fu, J., Vara Prasad, P.V. (2008). Heat-induced accumulation of chloroplast protein synthesis elongation factor, EF-Tu, in winter wheat. *J. Plant Physiol., 165*(2), 192-202.
[http://dx.doi.org/10.1016/j.jplph.2007.03.003] [PMID: 17498838]

Ristic, Z., Bukovnik, U., Prasad, P.V.V. (2007). Correlation between heat stability of thylakoid membranes and loss of chlorophyll in winter wheat under heat stress. *Crop Sci., 47*, 2067-2073.
[http://dx.doi.org/10.2135/cropsci2006.10.0674]

Saadalla, M.M., Quick, J.S., Shanahan, J.F. (1990). Heat tolerance in winter wheat: II. membrane thermostability and field performance. *Crop Sci., 30*, 1248-1251.
[http://dx.doi.org/10.2135/cropsci1990.0011183X003000060018x]

Safriel, U., Adeel, Z., Niemeijer, D., Puigdefabregas, J., White, R., Lal, R. (2005). *"Dryland systems," in Millennium Ecosystem Assessment: Ecosystems and Human Well-Being: Current State and Trends: Findings of the Condition and Trends Working Group.* (Vol. 1, pp. 623-662). Washington, DC: Island Press.

Saidi, Y., Peter, M., Finka, A., Cicekli, C., Vigh, L., Goloubinoff, P. (2010). Membrane lipid composition affects plant heat sensing and modulates $Ca^{2+}$-dependent heat shock response. *Plant Signal. Behav., 5*(12), 1530-1533.

[http://dx.doi.org/10.4161/psb.5.12.13163] [PMID: 21139423]

Sakai, A., Larcher, W. (1987). Frost survival of plants. *Responses and Adaptation to Freezing Stress..* Springer-Verlag.
[http://dx.doi.org/10.1007/978-3-642-71745-1]

Sanchez, Y., Lindquist, S.L. (1990). HSP104 required for induced thermotolerance. *Science,* *248*(4959), 1112-1115.
[http://dx.doi.org/10.1126/science.2188365] [PMID: 2188365]

Sánchez-Gómez, D., Valladares, F., Zavala, M.A. (2006). Performance of seedlings of Mediterranean woody species under experimental gradients of irradiance and water availability: Trade-offs and evidence for niche differentiation. *New Phytol.,* *170*(4), 795-806.
[http://dx.doi.org/10.1111/j.1469-8137.2006.01711.x] [PMID: 16684239]

Sangwan, V., Orvar, B.L., Beyerly, J., Hirt, H., Dhindsa, R.S. (2002). Opposite changes in membrane fluidity mimic cold and heat stress activation of distinct plant MAP kinase pathways. *Plant J.,* *31*(5), 629-638.
[http://dx.doi.org/10.1046/j.1365-313X.2002.01384.x] [PMID: 12207652]

Sanjana, P. (2015). Chapter 4. Genetic improvement in pearl millet. In: Tonapi, V.A., Patil, J.V., (Eds.), *Millets: Ensuring Climate Resilience and Nutritional Security.* (pp. 197-231). New Delhi: Astral publications, Daya Publishing house.

Schilmiller, A.L., Pichersky, E., Last, R.L. (2012). Taming the hydra of specialized metabolism: how systems biology and comparative approaches are revolutionizing plant biochemistry. *Curr. Opin. Plant Biol.,* *15*(3), 338-344.
[http://dx.doi.org/10.1016/j.pbi.2011.12.005] [PMID: 22244679]

Schoffl, F., Prandl, R., Reindl, A. (1999). Molecular responses to heat stress. In: Shinozaki, K., Yamaguchi-Shinozaki, K., (Eds.), *Molecular Responses to Cold, Drought, Heat and Salt Stress in Higher Plants.,* Austin, Texas: Landes Co. pp. 81-98.

Schöffl, F., Prändl, R., Reindl, A. (1998). Regulation of the heat-shock response. *Plant Physiol.,* *117*(4), 1135-1141.
[http://dx.doi.org/10.1104/pp.117.4.1135] [PMID: 9701569]

Senthil-Kumar, M., Kumar, G., Srikanthbabu, V., Udayakumar, M. (2007). Assessment of variability in acquired thermotolerance: potential option to study genotypic response and the relevance of stress genes. *J. Plant Physiol.,* *164*(2), 111-125.
[http://dx.doi.org/10.1016/j.jplph.2006.09.009] [PMID: 17207553]

Shah, F., Huang, J., Cui, K., Nie, L., Shah, T., Chen, C., Wang, K. (2011). Impact of high-temperature stress on rice plant and its traits related to tolerance. *J. Agric. Sci.,* *149*, 545-556.
[http://dx.doi.org/10.1017/S0021859611000360]

Shanahan, J.F., Edwards, I.B., Quick, J.S., Fenwick, J.R. (1990). Membrane thermostability and heat tolerance of spring wheat. *Crop Sci.,* *30*, 247-251.
[http://dx.doi.org/10.2135/cropsci1990.0011183X003000020001x]

Sharkey, T.D., Wiberley, A.E., Donohue, A.R. (2008). Isoprene emission from plants: why and how. *Ann. Bot.,* *101*(1), 5-18.
[http://dx.doi.org/10.1093/aob/mcm240] [PMID: 17921528]

Shi, Q., Bao, Z., Zhu, Z., Ying, Q., Qian, Q. (2006). Effects of different treatments of salicylic acid on heat tolerance, chlorophyll fluorescence, and antioxidant enzyme activity in seedlings of *Cucumis sativa* L. *Plant Growth Regul.,* *48*, 127-135.
[http://dx.doi.org/10.1007/s10725-005-5482-6]

Shirasawa, K., Sekii, T., Ogihara, Y., Yamada, T., Shirasawa, S., Kishitani, S., Sasaki, K., Nishimura, M., Nagano, K., Nishio, T. (2013). Identification of the chromosomal region responsible for Breeding Cultivars for Heat Stress Tolerance in Staple Food Crops high-temperature stress tolerance during the grain-filling period in rice. *Molecular Breeding,* *32*, 223-232.

[http://dx.doi.org/10.5772/intechopen.7648071]

Singh, B. (1993). Supra-optimal temperature tolerance in pearl millet inheritance pattern of some adaptive traits at seedling stage. *M.Sc. Thesis, CCS HAU, Hisar.*

Singh, R.V., Sharma, T.R., Khedar, O.P. (2003). Combining ability for seedling heat tolerance in pearl millet (*Pennisetum glaucum* (L.) R. Br.). *Indian J. Genet., 63*, 349. a

Singh, S., Nehra, D.S., Singh, R. (2003). *Haryana Agricultural-University-Journal-of-Research, 32*, 69-71. b

Smillie, R.M., Gibbens, G.C. (1981). Heat tolerance and heat hardening in crop plants measured by chlorophyll fluorescence. *Carlborg Res. Comm., 46*, 395-403. [http://dx.doi.org/10.1007/BF02907961]

Soman, P., Peacock, J.M. (1985). A laboratory technique to screen seedling emergence of sorghum and pearl millet at high soil temperature. *Exp. Agric., 21*, 335-341. [http://dx.doi.org/10.1017/S0014479700013168]

Soman, P., Peacock, J.M., Bidinger, F.R. (1984). A field technique to screen seedling emergence of pearl millet and sorghum through soil crust. *Exp. Agric., 20*, 327-334. [http://dx.doi.org/10.1017/S0014479700018019]

Squire, G.R. (1989). Response to temperature in a stand of pearl millet. *J. Exp. Bot., 40*(221), 1391-1398. [http://dx.doi.org/10.1093/jxb/40.12.1391]

Stomph, T.J. (1990). *Seedling Establishment in Pearl Millet [Pennisetum glaucum (L.) R. Br.]: the Influence of Genotype, Physiological Seed Quality Soil Temperature and Soil Water.* Ph. D. Thesis, University of Reading, Reading.

Sullivan, C.Y. (1972). Mechanisms of heat and drought resistance in grain sorghum and methods of measurement. *Sorghum in the seventies.* Oxford and IPH publishing house Co. New Delhi, India.

Swamy, P.M., Smith, B.N. (1999). Role of abscisic acid in plant stress tolerance. *Curr. Sci., 76*, 1220-1227.

Takeda, H., Cenpukdee, U., Chauhan, Y.S., Srinivasan, A., Hussain, M.M., Rashad, M.M., Lin, B.Q. (1999). Studies in Heat Tolerance of Brassica Vegetables and Legumes at the International Collaboration Research Section from 1992 to 1996. *Proceedings of a workshop on heat tolerance of crops,* Okinawa, Japan 17-29.

Talukder, S.K., Babar, M.A., Vijayalakshmi, K., Poland, J., Prasad, P.V.V., Bowden, R., Fritz, A. (2014). Mapping qtl for the traits associated with heat tolerance in wheat (*Triticum aestivum* L.). *BMC Genet., 15*, 97. [http://dx.doi.org/10.1186/s12863-014-0097-4] [PMID: 25384418]

Touchette, B.W., Smith, G.A., Rhodes, K.L., Poole, M. (2009). Tolerance and Avoidance: Two Contrasting Physiological Responses to Salt Stress in Mature Marsh Halophytes *Juncus roemerianus* Scheele and *Spartina alterniflora* Loisel. *J. Exp. Mar. Biol. Ecol., 380*, 106-112. [http://dx.doi.org/10.1016/j.jembe.2009.08.015]

Wahid, A.S., Gelani, M. (2007). Heat tolerance in plants: An overview. *Environ. Exp. Bot., 61*, 199-223. [http://dx.doi.org/10.1016/j.envexpbot.2007.05.011]

Wang, W., Vinocur, B., Shoseyov, O., Altman, A. (2004). Role of plant heat-shock proteins and molecular chaperones in the abiotic stress response. *Trends Plant Sci., 9*(5), 244-252. [http://dx.doi.org/10.1016/j.tplants.2004.03.006] [PMID: 15130550]

Wiegand, C., Namken, L. (1966). Influences of plant moisture stress, solar radiation, 773 and air temperature on cotton leaf temperature. *Agron. J., 58*, 582-586. [http://dx.doi.org/10.2134/agronj1966.00021962005800060009x]

Wolfe, M.D., Tonsor, S.J. (2014). Adaptation to spring heat and drought in northeastern Spanish Arabidopsis thaliana. *New Phytol., 201*(1), 323-334. [http://dx.doi.org/10.1111/nph.12485] [PMID: 24117851]

Xiong, L., Lee, H., Ishitani, M., Zhu, J.K. (2002). Regulation of osmotic stress-responsive gene expression by

the LOS6/ABA1 locus in Arabidopsis. *J. Biol. Chem., 277*(10), 8588-8596.
[http://dx.doi.org/10.1074/jbc.M109275200] [PMID: 11779861]

Yadav, A.K., Narwal, M.S., Singh, B. (2006). Field screening technique for heat effect on seedlings of pearl millet (Pennisetum glaucum L. R. Br.). *National Seminar on Transgenic Crops in Indian Agriculture: Status, Risk and Acceptance.* 28-29 January 2006, National Society of plant science (pp 109–12), CCS HAU, Hisar India.

Yadav, A.K., Arya, R.K., Narwal, M.S. (2014). Screening of pearl millet F1 hybrids for heat tolerance at early seedling stage. *Advances in Agriculture,* 1-17. Article ID 231301
[http://dx.doi.org/10.1155/2014/231301]

Yadav, A.K., Narwal, M.S., Arya, R.K. (2011). Genetic dissection of temperature tolerance in pearl millet (*Pennisetum glaucum*). *Indian J. Agric. Sci., 81*(3), 203-213.

Yadav, A.K., Narwal, M.S., Arya, R.K. (2012). Study of genetic architecture for maturity traits in relation to supra-optimal temperature tolerance in pearl millet (*Pennisetum glaucum* (L.) R. Br.). *Int. J. Plant Breed. Genet., 6*(3), 115-128.
[http://dx.doi.org/10.3923/ijpbg.2012.115.128]

Yadav, O.P., Bidinger, F.R. (2008). Dual-purpose landraces of pearl millet (*Pennisetum glaucum*) as sources of high stover and grain yield for arid zone environments. *Plant Genet. Resour., 6*(2), 73-78.
[http://dx.doi.org/10.1017/S1479262108993084]

Yadav, O.P., Weltzein, E., Bidinger, F.R. (2004). Diversity among pearl millet landraces collected in north western India. *Ann. Arid Zone, 43*(1), 45-53.

Yadav, A.K., Narwal, M.S., Arya, R.K. (2009). Screening for supraoptimal temperature tolerance through membrane thermo-stability in pearl millet (*Pennisetum glaucum*). *Forage Res., 35,* 85-90.

Yadav, A.K., Narwal, M.S., Arya, R.K. (2013). Evaluation of pearl millet (*Pennisetum glancum*) genotypes and validation of screening methods for supra-optimal temperature tolerance at seedling stage. *Indian J. Agric. Sci., 83*(3), 260-271.

Yang, J., Zhang, J. (2006). Grain filling of cereals under soil drying. *New Phytol., 169*(2), 223-236.
[http://dx.doi.org/10.1111/j.1469-8137.2005.01597.x] [PMID: 16411926]

Zeng, Y., Conner, J., Ozias-Akins, P. (2011). Identification of ovule transcripts from the Apospory-Specific Genomic Region (ASGR)-carrier chromosome. *BMC Genomics, 12,* 206.
[http://dx.doi.org/10.1186/1471-2164-12-206] [PMID: 21521529]

Zhang, N., Carlucci, P., Nguyen, J., Hayes-Jackson, J.W., Tonsor, S.J. (2016). Contrasting avoidance - tolerance in heat stress response from thermally contrasting climates in *Arabidopsis thaliana. bioRxiv.*
[http://dx.doi.org/10.1101/044461]

Zlatev, Z., Lidon, F., Ramalho, J., Yordanov, I. (2006). Comparison of resistance to drought of three bean cultivars. *Biol. Plant., 50,* 389-394.
[http://dx.doi.org/10.1007/s10535-006-0054-9]

# CHAPTER 4

# Advances In Breeding For Heat Stress Tolerance In Chickpea

**B. S. Patil**[1,*]**, Jayant S. Bhat**[1]**, A. G. Vijaykumar**[2]**, C. Bharadwaj**[3] **and U. C. Jha**[4]

[1] *Regional Centre, ICAR-Indian Agricultural Research Institute, Dharwad-580005, India*

[2] *Seed Unit, UAS Dharwad, India*

[3] *Division of Genetics, IARI New Delhi, India*

[4] *Indian Institute of Pulses Research, Kanpur (U.P.), India*

**Abstract:** High temperature stress is one of the important abiotic stresses hindering in achieving potential yield in crop plants, particularly cool-season grain legumes. Chickpea is one of the important cool-season grain legume crops. It experiences high temperature stress at different growth stages. Prevalence of heat stress during the reproductive stage reduces the crop yield drastically. Although genetic resource is available for heat stress tolerance in chickpea, studies on inheritance and its utilization in breeding program remain very limited. Research efforts through conventional breeding have been targeted to identify the traits for indirect selection. Advancement of molecular breeding approaches has led to the identification of markers linked to traits contributing to heat stress tolerance. Despite the availability of large scale genomic resources, most of the studies were limited to identify the molecular markers linked to quantitative trait loci (QTL). The functional genomics provides better insight into the molecular pathways and functions of the genes involved in heat stress tolerance. Limited information is available on the genes and pathways of gene activation controlling effective stress resistance in chickpea. Genome-wide analysis of *Hsfs* gene family resulted in the identification of *Hsf* genes which belong to four major groups with several paralogous and orthologous genes, and are unevenly distributed across all of the eight chromosomes. The next-generation sequencing and genome-editing techniques will greatly contribute in designing abiotic stress tolerant crop plants including chickpea.

**Keywords:** Chickpea, Genetic variability, Genomics, Heat stress.

## INTRODUCTION

Chickpea is a cool-season pulse crop, mainly cultivated in the rainfed ecology of

---

* **Corresponding author B. S. Patil**: Regional Centre, ICAR-Indian Agricultural Research Institute, Dharwad-580005, India; E-mail: bs_patil2000@yahoo.com

**Uday C. Jha, Harsh Nayyar and Sanjeev Gupta (Eds.)**
**All rights reserved-© 2020 Bentham Science Publishers**

the Indian subcontinent, the Mediterranean region, the West Asian and North American region, East-African region and Latin America (Rani *et al.*, 2020). Change in climatic conditions and changes in cropping systems are pushing chickpea cultivation to relatively warmer growing conditions. For example, in India, there has been a major shift in the chickpea area (about 3.0 million ha) from northern India (cooler, long season environment) to southern India (warmer, short season environment) during the past four decades. The major factor contributing to this shift is the change in cropping system replacing chickpea with wheat in Northern India. The cultivation of chickpea in warmer climatic conditions is facing new biotic and abiotic stresses. Among the abiotic stresses, high temperature stress is becoming a major challenge.

Globally, India is the largest producer of chickpea, accounting for 65% (9.075 million tonnes) of the total production. Australia is the second leading country with a 14% share in chickpea production (Merga and Haji, 2019). In both countries, chickpea is exposed to high temperature stress in the growing season, mainly during the reproductive phase (Devasirvatham, 2012a). Brief exposure of plants to high temperature stress during the reproductive phase can accelerate senescence, diminish seed set and seed weight, and ultimately, reduce yield (Siddique *et al.*, 1999).

The United Nations Inter Governmental Panel on Climate Change (IPCC) has projected an increase in global average temperature by 1.5 to 2.8 °C over the next century (Jones *et al.*, 1999). The Indian subcontinent and South Asian regions are believed to experience an increase in temperature of 0.5 °C and a warming of 2-4°C by the end of this century. The report also projects that by the end of 21$^{st}$ century, rainfall in India will increase by 15.40 percent, warming will be more pronounced in parts of North India and there will be an increase of 10 percent rainfall during *kharif*, while there is an uncertain prediction of rainfall and rise in temperature during *rabi*.

High temperature stress is defined as the rise in environmental temperature beyond a threshold level for a period of time sufficient to cause irreversible damage to plant growth and development. Base threshold temperatures vary with plant species, but for cool-season crops, 0°C is often the best-predicted base temperature (Miller *et al.*, 2001).

## EFFECT OF HEAT STRESS AT VARIOUS PLANT STAGES

The different developmental stages of chickpea ranging from seedling to grain filling stage are significantly affected by high temperature stress (Rani *et al.*, 2020). The crop growth stage and duration of occurrence of heat stress affect the crop duration and grain yield. The prevalence of hot and dry conditions (>30°C) results in forced maturity by reducing the crop duration (Summerfield *et al.*, 1984). Physiologically this can be interpreted as a reduction in the source as well as sink size and ultimately affects the grain yield. The temperature at seed germination, seedling establishment and reproductive stage are very critical for chickpea (Table **1**).

**Table 1. Optimum temperature and traits affected by heat stress during different crop stages of chickpea.**

| Growth Stage | Optimum Temperature | Heat Stress | Traits Affected | Reference |
|---|---|---|---|---|
| Germination | 31°C - 33°C | >42.5°C | Membrane injury<br>Lack of embryo growth | Ibrahim, (2011)<br>Covell *et al.*, (1986) |
| Vegetative growth | 20°C - 27°C | >30°C | Photosynthesis,<br>Transpiration<br>Forced maturity | Singh and Dhaliwal, (1972)<br>Summerfield *et al.*, (1984) |
| Reproductive period | 20°C – 26°C | >35°C | Pollen viability<br>Stigma receptivity<br>Number of pods per plant<br>Seed size | Devasirvatham *et al.*, (2013)<br>Kaushal *et al.*, (2013)<br>Summerfield *et al.*, (1984)<br>Munier-Jolain and Ney, (1998) |

High temperature stress at the time of sowing affects seed germination and seedling establishment. The seed germination is affected mainly due to hindrance in mobilization of cotyledon reserves required for embryo growth (Covell *et al.*, 1986). Genotypic variation was observed in chickpea affecting rate of germination under various temperature (Ellis *et al.*, 1986). Chickpea seed germination completely ceased when the temperature ranged between 45 to 48°C (Singh and Dhaliwal, 1972). The seedling growth and development is affected due to reduction of photosynthetic rates and increased transpiration rates under high temperature stress, resulting in reduced plant establishment in chickpea (Singh and Dhaliwal, 1972).

The reproductive phase in chickpea is known to be very sensitive to changes in environmental conditions, and exposure to heat stress at this stage leads to a reduction in seed yield (Summerfield *et al.*, 1984). The male (pollen, anthers) and female (stigma, style, ovary) reproductive parts of the flower are most sensitive to heat stress (Fig. **1**). The pollen viability, stigma receptivity and ovule viability are useful indicators of sensitivity to heat stress in chickpea. At the time of flower initiation, high temperature stress affects pollen development. The small, shrunken and empty pollen produced under high temperature stress affects their functionality. Due to abnormal development of pollen grains under stress conditions, viability is affected and/or sterile pollen grains are produced.

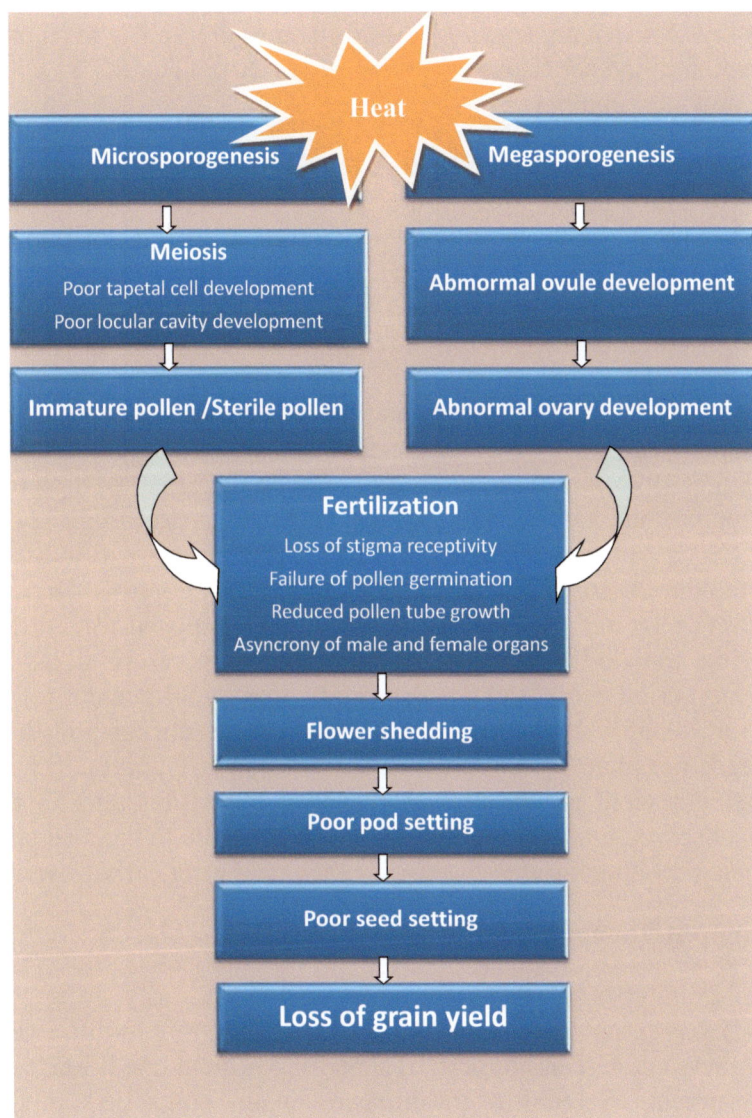

**Fig. (1).** Effect of heat stress during the reproductive phase in chickpea.

Thus, the failure of fertilization and flower drop is observed under heat stress. Heat stress immediately after flowering affects the germination and pollen tube growth in chickpea (Devasirvatham *et al.*, 2012b). At peak flowering and pod filling stage, exposure to higher temperature (>35°C) results in significant yield loss due to reduced seed size and seed number (Summerfield *et al.*, 1984; Wang *et al.*, 2006). The female reproductive part is also affected by heat stress. The failure of fertilization in chickpea by heat stress is due to reduced stigma receptivity (Kaushal *et al.*, 2013; Kumar *et al.*, 2013). The ovule and ovary abnormalities affect the seed and pod set (Devasirvatham *et al.*, 2013) leading to a reduction in seed weight and seed number.

The longer the period of exposure to high temperature stress during reproductive development to a high day temperature of 35°C, the lower the yield (Summerfield *et al.*, 1984). The majority of chickpea genotypes do not set pods when the temperature is > 35°C (Basu *et al.*, 2009). However, there is considerable genetic variation among chickpea genotypes for the response to heat stress. Pollen sterility is reported on exposure to high temperature (35/20 °C day/night) for 24 h before anthesis in the chickpea genotype ICC 5912, while fertility is retained in the genotype ICCV 92944 (Devasirvatham *et al.*, 2010). The reproductive period from anthesis to seed set is a highly critical stage for exposure to heat stress (Gross and Kigel 1994). Wang *et al.* (2006) studied the effect of high temperature stress on effective pod setting, seed setting and seed yield of chickpea. 'Myles' (*desi*) and 'Xena' (*kabuli*) chickpea varieties were grown in a controlled environment under 20/16 °C day/night air temperatures (control). High (35/16 °C) and moderate (28/16 °C) temperature stresses were imposed for 10 days during early flowering and pod development. The high temperature stress during the early flowering period decreased the pod production by 34 percent for Myles and 22 percent for Xena, whereas high temperature stress during pod development decreased seeds per plant by 33 percent for Myles and 39 percent for Xena. As a result, 59 percent yield reduction was recorded for Myles and 53 percent for Xena. It was observed that yield reduction was higher due to the heat stress during pod development as compared to the stress received during early flowering.

**High Throughput Phenotyping For Identifying Heat Tolerant Genotype**

Screening of genotypes against heat stress can be broadly grouped into two categories. (i) screening under controlled environmental condition (Phytotrons, Growth chambers and Green house). (ii) screening in field condition, representing target environment. Screening germplasm under controlled environmental condition provides an ideal condition to study physiological and molecular mechanisms governing heat stress tolerance. However, the validation has to be carried out under field conditions to confirm the results. The screening of chickpea genotypes against heat-stress during reproductive stage can be conducted by delaying the sowing date so as to coincide the reproductive phase of the crop with higher temperatures (Gaur *et al.*, 2014). A rapid and effective pollen screening technique for heat stress tolerance has been standardized by

Singh *et al.*, (2016). The maize pollen grains were incubated at a high temperature (36 °C) in a thermal cycler for 2-4 hrs. The heat treated pollen grains of maize inbred lines displayed differential response for *in vitro* pollen germination and capacity to fertilize and produce seed under *in vivo* fertilization. Normally under controlled condition, heat stress effect is studied at seedling stage. Earnest *et al.* (2017) studied the heat stress effect on pollen viability, pollen production and fertilization in lima bean by exposing the lima bean plants to high day/night temperature under controlled condition. A robust field based 'heat tent and cyber-physical system technology' to screen wheat lines for high night temperature stress was developed by Hein *et al.* (2019). This is a reliable, simple and cost-effective method. It provides a realistic field conditions to systematically record the observations on agronomic and physiological parameters in response to high night temperature. This method can be used to phenotype other crops as well. Recent advances in imaging technologies provide an efficient tool for precision phenotyping of heat-stress related traits. This allows rapid and non-destructive method of measuring heat stress responsive traits. The imaging based technology includes easy-to-use tools of spectral reflectance, digital imagery, thermal imagery, and stable isotopes. These tools could facilitate large-scale phenotyping of morphological and physiological traits in crop plants. However, precise, accurate and high-throughput phenotyping technique to screen large number of genotypes is yet to be developed. In crops like wheat and chrysanthemum, infrared thermography is used for large scale phenotyping against heat stress (Prashar and Jones, 2014). The new generation phenotyping platforms *viz.*, HTpheno (Hartmann *et al.*, 2011). and Rootscope (Kast *et al.*,2013) are being routinely used to screen the crop species against heat stress. The research efforts directed to develop cost-effective and user-friendly phenotyping platforms may allow to screen large number of germplasm against heat stress.

**Genetic Resources For Heat Stress Tolerance**

Remarkable efforts have been made to identify the genetic resource to various abiotic stresses in chickpea. Several chickpea genotypes (*e.g.*, ICC1052 and ICC 8522) avoiding heat stress by virtue of displaying high leaf area index (LAI) value have also been reported (Choudhary *et al.*, 2018). A heat-tolerant line ICCV 92944 has been released in three countries (as Yezin 6 in Myanmar and JG 14 in India and Chinadesi 2 in Kenya). Early flowering with quick canopy growth enables the chickpea genotype 'ICCV 92944' to perform well under late-sown conditions and thus, allow escaping heat stress (≥35 °C). Dua (2001) identified two heat-tolerant genotypes (ICCV 88512 and ICCV 88513) from the tested 25 chickpea genotypes. Based on heat tolerance index (HTI), three chickpea genotypes *viz.,* ICC 3362, ICC 6874 and ICC 12155 were identified as heat tolerant lines from 280 diverse chickpea lines screened for heat tolerance in two locations (Patancheru and Kanpur) (Krishnamurthy *et al.*, 2011). Upadhyaya *et al.* (2011) identified ICC 14346 as a heat-tolerant genotype from a set of 35 early maturing germplasm thus, this genotype could be potentially used as a donor parent in chickpea heat tolerance research. Based on pod setting ability of a genotype as selection criteria under high temperature of ≥37°C, Devasirvatham *et al.* (2012a, 2013) identified two chickpea genotypes ICC1205 and ICC15614 as heat-tolerant lines. In India, National Agricultural Research System (NARS) has developed heat-tolerant, short-duration varieties (JG 14, JSC 55, JSC 56) suitable for cultivation under different agro-climatic regions of the country. Heat stress tolerance is a highly complex trait. It involves heat stress induced expression of a cascade of genes followed by activation of associated complex molecular networks *viz.*, signal transduction, production of metabolites, heat shock proteins (Nadeem *et al.*, 2018). As far as genetics of heat stress tolerance in chickpea is concerned limited information is available. Polygenic inheritance of heat tolerance was reported by Upadhyaya *et al.*, (2011), and it is highly influenced by genotype × environment (G × E) interaction. Devasirvatham (2012a) opined that the pattern of inheritance of heat tolerance in chickpea is complex. Presence of both additive and non-additive genetic variance is reported by Jha *et al.* (2019a). The study also revealed presence of higher magnitude of GCA variance than SCA variance for most of the traits studied under heat stress condition, ICC 92944 and KWR 108 were better general combiners for yield and yield- related traits. Crop wild relatives (CWRs) are one of the potential sources of resistance genes for biotic and abiotic stresses. The *Cicer* genus has 9 annual and 34 perennial wild species. High cross compatibility of cultivated species was observed only with *C. reticulatum*. This limits the use of wild species in chickpea improvement programmes. Canci and Toker (2009) identified four accessions of *C. reticulatum* (AWC 605, AWC 616, AWC 620 and AWC 625) and one accession of *C. pinnatifidum* (AWC 500) as a source of resistant to drought and heat stress. The accessions belonging to other wild species either failed to germinate or found highly susceptible to drought and heat stress. Further, screening of large number of CWRs and landraces could help in identifying heat tolerance relevant traits. Thus, transferring of these traits in elite yet heat sensitive chickpea cultivars could help in sustaining chickpea yield under high temperature stress.

## Breeding Approaches For Designing Heat Stress Tolerance Crop Plant

In legumes, breeding for abiotic stress tolerance received less attention compared to that for biotic stress tolerance. Large genetic variability for heat tolerance exists in legumes (Kumar *et al.*, 2016; Bindumadhava *et al.*, 2017, Jha *et al.*, 2018). Conventional breeding approaches have been employed in development and release of many heat stress tolerant genotypes of legumes. In chickpea heat stress tolerant genotypes are developed by using phenotypic selection under field condition, and by using various heat tolerance indices (Gaur *et al.*, 2008; Krishnamurthy *et al.*, 2011). Jha *et al.* (2018) used different heat stress tolerance indices, *viz.*, mean productivity (MP), geometric mean productivity (GMP), yield index (YI), tolerance index (TOL), stress susceptibility index (SSI) and superiority measure (SM) to screen chickpea genotypes. Further they suggested four parameters (MP, YI, GMP and SSI) as most efficient selection indices to identify heat stress tolerant lines in chickpea. In other pulse crops such as common bean, lentil, mungbean, pea, groundnut and cowpea, improved varieties have been developed for cultivation under elevated temperature condition. In these crops the selection indices used include, stress tolerant index (STI), seedling vigour, early maturity, temperature induction response, solute leakage and chlorophyll fluorescence (Porch, 2006, Sultana *et al.*, 2014 and Bindumadava *et al.*, 2017). The most frequently used breeding methods include, recurrent selection, diverse cross combination, backcrossing, and pedigree method (Patel and Hall, 1990; Hall, 1992, 1993, 2011; Craufurd *et al.*, 2003; Lucas *et al.*, 2013). Many heat tolerant chickpea cultivars developed/identified through conventional breeding have been mentioned in other section.

## Traits Contributing For Heat Stress Tolerance

Identification of traits contributing for heat stress tolerance and robust screening techniques are pre-requisites in breeding for heat stress tolerance in chickpea. Screening under field conditions is very much complex as the traits contributing for heat tolerance have interaction with many environmental factors. On the other hand, screening under controlled environmental conditions has its own limitation of absence of environmental factors that interact with the mechanism governing the heat stress tolerance under natural environmental conditions. This leads to failure of lines selected under controlled environmental conditions to perform under field condition. Nevertheless, different physiological, phenological and biochemical traits are being used as selection criteria in breeding for heat stress tolerance in crop plants.

High temperature stress severely affects cell membrane structure and function (Wahid *et al.*, 2007). Damage due to heat stress can be determined by loss of membrane integrity (Salvucci and Crafts-Brandner, 2004). Therefore, membrane thermostability (MTS) as measure of electrolyte leakage due to membrane damage has been used to screen the genotypes under high-temperature stress conditions (Liu and Huang, 2000; Xu *et al.*, 2006). Membrane injury due to heat stress is reported in chickpea at 40/30°C, which was further intensified at 45/35°C (Kumar *et al.*, 2013). Based on MTS chickpea is considered as the most heat sensitive legume compared with other grain legumes such as pigeon pea, groundnut, and soybean (Devasirvatham *et al.*, 2012b).

Heat stress reduces the photosynthetic rate by decreasing leaf chlorophyll and nitrogen contents. Among the three major heat-sensitive sites in photosynthetic machinery, photosystem II (PSII) is highly sensitive in chickpea (Kaushal *et al.*, 2016). In chickpea, high temperature stress causes damage to chloroplast membranes by deterioration of photosynthetic pigments. Consequently, reduction in chlorophyll under high temperature has been reported in chickpea (Kumar *et al.*, 2013). The combination of these two physiological traits (MTS and chlorophyll content) is found to be highly reliable selection index to screen large number of genotypes in chickpea.

Heat stress is associated with increased transpiration (Kolb and Robberecht, 1996; Hasanuzzaman *et al.*, 2013). Increased transpiration results in loss of moisture from plants and reduction in reduced turgor pressure (Tsukaguchi *et al.*, 2003). In heat tolerant chickpea genotypes high stomatal conductance is recorded under heat stress. This enhances transpirational heat dissipation in tolerant chickpea genotypes as long as soil water is available (Kaushal *et al.*, 2013). Therefore, maintenance of leaf osmotic potential may also be considered as one of the physiological traits under heat stress to select tolerant lines in chickpea.

## Genomic Resources For Heat Stress Tolerance

Increasing chickpea genomic resources could provide better insight into the genetics of heat tolerance in crop plants. More than 3,000 simple sequence repeats (SSRs), diversity array technology (DArT) arrays (Thudi *et al.*, 2011) and huge number of single nucleotide polymorphism (SNP) markers developed by next generation sequencing technology (Hiremath *et al.*, 2012; Varshney *et al.*, 2013) have enriched chickpea genomic resources. Despite the availability of large scale genomic resources, most of the studies were limited to identify the molecular markers linked to quantitative trait loci (QTL). Majority of QTL studies were mostly limited to biotic stresses like Fusarium wilt, Ascochyta blight and Botrytis gray mold. (Thudi *et al.*, 2017). Very few studies were conducted to unearth the genomic resources for heat stress tolerance in chickpea (Table **2**). However, availability of DArT marker allowed conducting association mapping to identify several significant marker traits association for 11 traits under normal and heat stress in chickpea (Kaloki *et al.*, 2019). Further, genotyping by sequencing (GBS)-based single nucleotide polymorphism markers were used to identify genomic regions responsible for heat tolerance in chickpea (Paul *et al.*, 2018a).

**Table 2. 'Omics' resources for heat stress tolerance in chickpea.**

| Genomic Resources | | | | | |
|---|---|---|---|---|---|
| **Type of Mapping Population** | **Name of QTLs** | **Type of Markers** | **LG Groups** | **PV%** | **References** |
| DCP92-3 × ICC92944 ($F_2$, 206) | One QTL related to chlorophyll content, One QTL controlling primary branch number | SSR | LG3 and LG6 | 2-17.2 | Jha *et al.* (2019b) |
| ICC 15614 × ICC 4567 | *qfpod02_5* *qfpod03_6* *qgy02_5* *qgy03_6* | SNP | LG05, LG06 | 6.5-11.5 | Paul *et al.* (2018b) |
| Association panel (300) | 312 significant MTAs | SSR and SNP | - | - | Thudi *et al.* (2017) |
| Association panel (71) | Several significant MTAs for membrane stability and chlorophyll content | SSR | LG1, 2, 3,4,5.6, and 7 | 5.5-22 | Jha *et al.* (2018) |
| Transcriptomic Resources | | | | | |
| **Name of Genotype** | **Candidate Genes** | **Function** | **Plateform Used** | | **Reference** |
| Three tolerant (ICCV 92944, ICC 1356, ICC 15614) Three sensitive (ICC 5912, ICC 4567, ICC 10685) | *Ca_25811, Ca_23016, Ca_09743, Ca_17680* and *Ca_25602* | Contribute to heat stress tolerance | RNA-seq qRT-PCR | | Agarwal *et al.* (2016) |

Eight QTLs for heat tolerance in chickpea were identified using 292 $F_{8:9}$ recombinant inbred lines (RILs) developed from the cross, ICC 4567 (heat sensitive) × ICC 15614 (heat tolerant). Four major QTLs governing number of filled pods per plot, and grain yield per plot located on the CaLG05, and CaLG06 were uncovered from ICC15614 x ICC4567 mapping population. Considering seed weight trait Thudi *et al.* (2017) elucidated 70 significant markers linked with this trait under drought and heat stresses in chickpea (Thudi *et al.*, 2017). Among the different physiological traits used for measuring high temperature tolerance, membrane stability index (MSI) and leaf chlorophyll content were studied by Jha *et al.* (2018). Association analysis using 71 chickpea genotypes resulted in identification of two markers (NCPGR206 and H2L102) linked with the MSI trait. Three SSR markers GA9, TR31 and TA113 showed significant association with chlorophyll content. Thus, further deployment of chickpea genomic resource could allow identifying more heat tolerant related QTLs and the underlying candidate genes contributing heat tolerance.

## Functional Genomics For Heat Stress Tolerance

Current advances in functional genomics have provided us great opportunity to underpin the possible candidate gene(s) and their function in controlling heat tolerance in crop plants (Jha *et al.*, 2014). Functional genomics has been widely used for studying the stress responses in different crop species, such as tomato (Gibly *et al.*, 2004), rice (Fujiwara *et al.*, 2004), maize (Baldwin 1998), cassava (Lopez *et al.*, 2005), soybean (Moy *et al.*, 2004) and *Arabidopsis thaliana* (Huitema *et al.*, 2003). The functional genomics provides better insight into the molecular pathways and functions of the genes involved heat stress tolerance. Various functional genomic approaches ranging from microarray based techniques, suppression subtractive hybridization (SSH) to RNA-seq techniques have been devoted to investigate the function of genes controlling heat tolerance (Zhang *et al.* 2005; Bita *et al.*, 2011; Agarwal *et al.*, 2016). However, limited functional genomics resources for heat tolerance in chickpea is available. Recently, Agarwal *et al.*(2016) uncovered several important candidate genes such as *Ca_25811, Ca_23016, Ca_09743, Ca_17680* and *Ca_25602* those involved in contributing heat tolerance through transcriptome sequencing of heat tolerant and heat sensitive contrasting parents in chickpea. Under high temperature stress, plants activate thermo-stress responsive pathways to rapidly accumulate heat shock proteins (HSPs). These HSPs play a major role in protecting cells against stress damage, folding; intracellular distribution; degradation of proteins, and in signal transduction pathways (Hartl *et al.*, 2002; Young *et al.*, 2003). Heat shock proteins are produced by several heat shock genes. The heat shock factor binding proteins (HSBPs) control the expression of these genes (Chen and Zhang, 1997). The heat shock transcription factors (HSFs) regulate the expression of heat shock genes by recognizing the conserved binding motifs (heat stress element, HSE) which is located in their promoter region. A genome-wide analysis of *Hsfs* gene family resulted in the identification of 21 *Hsf* genes in chickpea, of which only *PIE1* gene is upregulated during heat stress at pod development stage (Parameswaran *et al.*, 2016). Similarly, Zafar *et al.* (2016) characterized HSF family in chickpea. HSF genes in chickpea belong to four major groups with several paralogous and orthologous genes and are unevenly distributed across all of the eight chromosomes. Segmental duplications are the major cause for HSF gene family expansion during evolution.

WRKY proteins play a vital role in the regulation of several plant metabolic processes and pathways under biotic and abiotic stress condition. Waqas *et al.* (2019) identified 70 WRKY-encoding non-redundant genes in chickpea. The transcriptome data-based *in silico* expression analysis revealed that some of these genes have identical expression pattern under different stresses, revealing the possibility of involvement of these genes in conserved abiotic stress–response pathways. Parankusum *et al.* (2017) identified a set of 482 heat-responsive proteins in the tolerant genotype using comparative gel-free proteomics. Besides heat shock proteins, proteins such as acetyl-CoA carboxylase, pyrroline-5-carboxylate synthase (P5CS), ribulose-1,5-bisphosphate carboxylase/oxygenase (RuBisCO), phenylalanine ammonia-lyase (PAL) 2, ATP synthase, glycosyl transferase, sucrose synthase and late embryogenesis abundant (LEA) proteins were reported to be strongly associated with heat tolerance in chickpea.

**Prospects Of Innovative Breeding Approaches For Heat Stress Tolerance In Crop Plant**

The simultaneous developments in genotyping and phenotyping platforms have paved the way for better understanding of heat stress tolerance in crop plants. The other dimension in omics era is reverse genetics that provides an opportunity for genome editing to enhance heat stress tolerance in crop plants. There are three main gene editing techniques (1) Transcription Activator-Like Effector Nucleases (TALENs) (2) Zinc Finger Nucleases (ZFNs) and, (3) Clustered Regularly Interspaced Short Palindromic Repeats (CRISPR). These can be used to study the function of the gene and to modify them to enhance its function. The CRISPR is a simpler, accurate and faster genome editing technique when compared to other two techniques. Although this technique has been widely used in plant science, very few reports on its application for understanding and development of abiotic stress resistant plants are available (Vats *et al.*, 2019). In tomato, improved fruit setting under heat stress is achieved by CRISPR/Cas mediated genome editing of an 'S' gene, SlAGAMOUS-LIKE 6 (SlAGL 6) (Klap *et al.*, 2017). The CRISPR-Cas9 technique was efficiently used to understand the function of OsAnn3 gene in cold stress tolerance in rice (Shen *et al.*, 2017). In maize drought tolerant variety was developed by ARGOS8 gene editing through CRISPR/Cas based genome editing approach (Shi *et al.*, 2017). Likewise, two genes OsRR22 and OsNAC041 have been targeted for improvement of salinity tolerance in rice (Zhang *et al.*, 2019 and Bo *et al.*, 2019). A recent approach of multiplexed genome editing can be helpful for simultaneous targeting of multiple genes involved in abiotic stress tolerance (Vats *et al.*, 2019). The two categories of genes which could be targeted for gene editing are, regulatory and structural genes (Zafar *et al.*, 2019). The genes encoding antioxidant enzymes, which in turn scavenge the excessive reactive oxygen species and contribute for abiotic stress tolerance are termed as 'Tolerance genes' (T gene). The genes responsible for excessive production of ROS are termed as 'Sensitive genes' (S gene). The CRISPR/Cas has potential application in knocking out S genes in otherwise high yielding cultivars and over expression of 'T' genes by editing regulatory genes for enhanced tolerance to abiotic stresses. Among the two approaches, silencing of 'S' genes holds upper hand.

## SUMMARY AND THE FUTURE PROSPECTS

The heat stress can be studied using a holistic approach that integrates, physiological and biochemical characterization of plant responses to help define plant breeding strategies. These combined approaches also encompass molecular tools and agronomic practices that will have significant role in developing heat-tolerant chickpea cultivars. Although both plant male and female reproductive organ development processes are affected by temperature stress, development of male plant reproductive organ is more sensitive than that of female plant reproductive organ development. Carbon starvation due to reduced supply of photosynthates and carbohydrate diversion for antioxidants, accumulation of heat shock proteins and osmolytes, are proposed to be major contributors to reproductive heat response and adaptation. Understanding the molecular basis of heat stress tolerance and identification of QTLs will accelerate the breeding for reproductive thermo-tolerance. The other research gaps that need attention for developing heat tolerant chickpea cultivars are; development of simple and efficient screening technique to identify heat tolerant germplasm; large number of genotypes/breeding lines can be phenotyped with ease. Different traits contributing towards heat stress tolerance at different stages of crop growth and establishment of relationship among these traits help in devising appropriate breeding strategies. Understanding the physiological response to heat stress and determining the underlying genetic control of these responses.

With the advent of plant functional genomics tools, many novel genes related to thermo-tolerance have been identified and are being used to improve heat stress tolerance with the help of innovative approaches. Next generation sequencing and genome-editing techniques will play a vital role in crop improvement. The advanced tools will provide novel insights into heat tolerance mechanisms and help in better designing of heat tolerant high yielding climate smart chickpea varieties.

## CONSENT FOR PUBLICATION

Not applicable.

## CONFLICT OF INTEREST

The authors confirm that this chapter content has no conflict of interest.

## ACKNOWLEDGEMENTS

We thank all the senior colleagues and the anonymous reviewers for suggestions that were incorporated to improve this manuscript.

## REFERENCES

Agarwal, G., Garg, V., Kudapa, H., Doddamani, D., Pazhamala, L.T., Khan, A.W., Thudi, M., Lee, S.H., Varshney, R.K. (2016). Genome-wide dissection of AP2/ERF and HSP90 gene families in five legumes and expression profiles in chickpea and pigeonpea. *Plant Biotechnol. J., 14*(7), 1563-1577.
[http://dx.doi.org/10.1111/pbi.12520] [PMID: 26800652]

Baldwin, I.T. (1998). Jasmonate-induced responses are costly but benefit plants under attack in native populations. *Proc. Natl. Acad. Sci. USA, 95*(14), 8113-8118.
[http://dx.doi.org/10.1073/pnas.95.14.8113] [PMID: 9653149]

Basu, P.S., Ali, M., Chaturvedi, S.K. (2009). Terminal heat stress adversely affects chickpea productivity in Northern India – Strategies to improve thermotolerance in the crop under climate change. *In 'ISPRS Archieves XXXVIII-8/W3 Workshop Proceedings: Impact of Climate Change on Agriculture'*. 23-25 Feb. 2009 New Delhi, India 189-193.

Bindumadhava, H., Nair, R.M., Nayyar, H., Riley, J.J., Easdown, W. (2017). Mungbean production under a changing climate-insights from growth physiology. *Mysore J. Agric. Sci., 51*, 21-26.

Bita, C.E., Zenoni, S., Vriezen, W.H., Mariani, C., Pezzotti, M., Gerats, T. (2011). Temperature stress differentially modulates transcription in meiotic anthers of heat-tolerant and heat-sensitive tomato plants. *BMC Genomics, 12*, 384.
[http://dx.doi.org/10.1186/1471-2164-12-384] [PMID: 21801454]

Canci, H., Toker, C. (2009). Evaluation of annual wild *Cicer* species for drought and heat resistance under field conditions. *Genet. Resour. Crop Evol., 56*, 1-6.
[http://dx.doi.org/10.1007/s10722-008-9335-9]

Chen, J.N., Zhang, X.T. (1997). New progress in research on fuctions of heat shock protein in human and plants. *Hereditas, 19*, 45-48.

Choudhary, A. K. (2018). Integrated physiological and molecular approaches to improvement of abiotic stress tolerance in two pulse crops of the semi-arid tropics. *Crop J., 6*, 99-114.
[http://dx.doi.org/10.1016/j.cj.2017.11.002]

Covell, S., Ellis, R.H., Roberts, E.H. (1986). Summerfield RJ. The influence of temperature on seed germination rate in grain legumes I. A comparison of chickpea, lentil, soybean, and cowpea at constant temperatures. *J. Exp. Bot., 37*, 705-715.
[http://dx.doi.org/10.1093/jxb/37.5.705]

Craufurd, P.Q., Prasad, P.V.V., Kakani, V.G., Wheeler, T.R., Nigam, S.N. (2003). Heat tolerance in groundnut. *Field Crops Res., 80*, 63-77.
[http://dx.doi.org/10.1016/S0378-4290(02)00155-7]

Devasirvatham, V., Tan, D.K.Y., Gaur, P.M., Raju, T.N., Trethowan, R.M. (2012). High temperature tolerance in chickpea and its implications for plant improvement. *Crop Pasture Sci., 63*, 419-428. a
[http://dx.doi.org/10.1071/CP11218]

Devasirvatham, V., Tan, D.K.Y., Trethowan, R.M., Gaur, P.M., Mallikarjuna, N. (2010). Impact of high temperature on the reproductive stage of chickpea. In Food Security from Sustainable Agriculture Proceedings of the 15th Australian Society of Agronomy Conference 15-18.

Devasirvatham, V., Gaur, P.M., Mallikarjuna, N., Tokachichu, R.N., Trethowan, R.M., Tan, D.K.Y. (2012). Effect of high tmperature on the reproductive development of chickpea genotypes under controlled environments. *Funct. Plant Biol., 39*(12), 1009-1018.

[http://dx.doi.org/10.1071/FP12033] [PMID: 32480850]

Devasirvatham, V., Gaur, P.M., Mallikarjuna, N., Raju, T.N., Trethowan, R.M., Tan, D.K. (2013). Reproductive biology of chickpea response to heat stress in the field is associated with the performance in controlled environments. *Field Crops Res., 142*, 9-19.
[http://dx.doi.org/10.1016/j.fcr.2012.11.011]

Dixit, G.P., Srivastava, A.K. (2019). Singh NP Marching towards self-sufficiency in chickpea. *Curr. Sci., 116*(2), 239-242.
[http://dx.doi.org/10.18520/cs/v116/i2/239-242]

Dua, R.P. (2001). Genotypic variations for low and high temperature tolerance in gram (*Cicer arietinum*). *Indian J. Agric. Sci., 71*, 561-566.

Ellis, R.H., Covell, S., Roberts, E.H., Summerfield, R.J. (1986). The influence of temperature on seed germination rate in grain legumes. *J. Exp. Bot., 37*, 1503-1515.
[http://dx.doi.org/10.1093/jxb/37.10.1503]

Ernest, E. G., Wisser, R. J., Johnson, G. C. (2017). Physiological Effects of Heat Stress on Lima Bean (Phaseolus lunatus) and Development of Heat Tolerance Screening Techniques. *Physiological report of the bean improvement cooperative, 60*, 101-102.

Fujiwara, S., Tanaka, N., Kaneda, T., Takayama, S., Isogai, A., Che, F.S. (2004). Rice cDNA microarray-based gene expression profiling of the response to flagellin perception in cultured rice cells. *Mol. Plant Microbe Interact., 17*(9), 986-998.
[http://dx.doi.org/10.1094/MPMI.2004.17.9.986] [PMID: 15384489]

Gaur, P.M., Jukanti, A.K., Samineni, S., Chaturvedi, S.K., Basu, P.S., Babbar, A., Jayalakshami, V., Nayyar, H., Devasirvatham, V., Mallikarjuna, N., Krishnamurthy, L., Gowda, C.L.L. (2004). *Climate Change and Heat Stress Tolerance in Chickpea.* (pp. 839-850). KGaA, Boschstr 12, 69469 Weinheim, Germany: Wiley-VCH Verlag GmbH & Co.

Gaur, P.M., Krishnamurthy, L., Kashiwagi, J. (2008). Improvement of drought-avoidance root traits in chickpea (*Cicer arietinum* L.) –Current status of research at ICRISAT. *Plant Prod. Sci., 1*, 3-11.
[http://dx.doi.org/10.1626/pps.11.3]

Gibly, A., Bonshtien, A., Balaji, V., Debbie, P., Martin, G.B., Sessa, G. (2004). Identification and expression profiling of tomato genes differentially regulated during a resistance response to *Xanthomonas campestris* pv. *vesicatoria. Mol. Plant Microbe Interact., 17*(11), 1212-1222.
[http://dx.doi.org/10.1094/MPMI.2004.17.11.1212] [PMID: 15553246]

Gross, Y., Kigel, J. (1994). Differential sensitivity to high temperature of stages in the reproductive development in common bean (*Phaseolus vulgaris* L.). *Field Crops Res., 36*, 201-212.
[http://dx.doi.org/10.1016/0378-4290(94)90112-0]

Hall, A.E. (2011). Breeding cowpea for future climates. In: Yadav, S.S., Redden, R.J., Hatfield, J.L., Lotze-Campen, H., (Eds.), *Crop Adaptation to Climate Change.* 340-355.
[http://dx.doi.org/10.1002/9780470960929.ch24]

Hall, A.E. (1992). Breeding for heat tolerance. *Plant Breed. Rev., 10*, 129-168.

Hall, A.E. (1993). Physiology and Breeding for Heat Tolerance in Cowpea and Comparisons with Other Crops. In: Kuo, C.G., (Ed.), *Adaptation of Food Crops to Temperature and Water Stress.* (pp. 271-284). Taipei, Taiwan: Asian Vegetable Research and Development Center.

Hartl, F.U., Hayer-Hartl, M. (2002). Molecular chaperones in the cytosol: from nascent chain to folded protein. *Science, 295*(5561), 1852-1858.
[http://dx.doi.org/10.1126/science.1068408] [PMID: 11884745]

Hartmann, A., Czauderna, T., Hoffmann, R., Stein, N., Schreiber, F. (2011). HTPheno: an image analysis pipeline for high-throughput plant phenotyping. *BMC Bioinformatics, 12*, 148.
[http://dx.doi.org/10.1186/1471-2105-12-148] [PMID: 21569390]

Hasanuzzaman, M., Nahar, K., Alam, M.M., Roychowdhury, R., Fujita, M. (2013). Physiological, biochemical, and molecular mechanisms of heat stress tolerance in plants. *Int. J. Mol. Sci., 14*(5), 9643-9684.
[http://dx.doi.org/10.3390/ijms14059643] [PMID: 23644891]

Hein, N.T., Wagner, D., Bheemanahalli, R., Šebela, D., Bustamante, C., Chiluwal, A., Neilsen, M.L., Jagadish, S.V.K. (2019). Integrating Field-based Heat Tents and Cyber-physical System Technology to Phenotype High Night-time Temperature Impact on Winter Wheat. *Plant Methods, 15*, 41.
[http://dx.doi.org/10.1186/s13007-019-0424-x] [PMID: 31044000]

Hiremath, P.J., Kumar, A., Penmetsa, R.V., Farmer, A., Schlueter, J.A., Chamarthi, S.K., Whaley, A.M., Carrasquilla-Garcia, N., Gaur, P.M., Upadhyaya, H.D., Kavi Kishor, P.B., Shah, T.M., Cook, D.R., Varshney, R.K. (2012). Large-scale development of cost-effective SNP marker assays for diversity assessment and genetic mapping in chickpea and comparative mapping in legumes. *Plant Biotechnol. J., 10*(6), 716-732.
[http://dx.doi.org/10.1111/j.1467-7652.2012.00710.x] [PMID: 22703242]

Huitema, E., Vleeshouwers, V.G., Francis, D.M., Kamoun, S. (2003). Active defence responses associated with non-host resistance of *Arabidopsis thaliana* to the oomycete pathogen *Phytophthora infestans. Mol. Plant Pathol., 4*(6), 487-500.
[http://dx.doi.org/10.1046/j.1364-3703.2003.00195.x] [PMID: 20569408]

Ibrahim, H.M. (2011). Heat stress in food legumes: evaluation of membrane thermostability methodology and use of infra-red thermostability. *Euphytica, 180*, 99-105.
[http://dx.doi.org/10.1007/s10681-011-0443-9]

Jha, U.C., Bohra, A., Singh, N.P. (2014). Heat stress in crop plants: its nature, impacts and integrated breeding strategies to improve heat tolerance. *Plant Breed., 133*, 679-701. b
[http://dx.doi.org/10.1111/pbr.12217]

Jha, U. C., Jha, R., Bohra, A., Parida, S. K., Kole, P. C., Thakro, V., Singh, D., Singh, N. P. (2018). Population structure and association analysis of heat stress relevant traits in chickpea (*Cicer arietinum L.*). *3 Biotech, 8*, 1-14.

Jha, U.C., Kole, P.C., Singh, N.P. (2019). Nature of gene action and combining ability analysis of yield and yield related traits in chickpea (*Cicer arietinum* L.) under heat stress. *Indian J. Agric. Sci., 89*, 500-508. a

Jha, U.C., Kole, P.C., Singh, N.P. (2019). QTL mapping for heat stress tolerance in chickpea (*Cicer arietinum* L.). *Legume Res.* b
[http://dx.doi.org/10.18805/LR-4121]

Jones, P.D., New, M., Parker, D.E., Mortin, S., Rigor, I.G. (1999). Surface area temperature and its change over the past 150 years. *Rev. Geophys., 37*, 173-199.
[http://dx.doi.org/10.1029/1999RG900002]

Kaloki, P., Devasirvatham, V., Tan, D.K.Y. (2018). Chickpea abiotic stresses: Combating drought, heat and cold.
[http://dx.doi.org/10.5772/intechopen] [PMID: 83404]

Kast, E.J., Nguyen, M.D.T., Lawrence, R.E., Rabeler, C., Kaplinsky, N.J. (2013). The RootScope: a simple high-throughput screening system for quantitating gene expression dynamics in plant roots. *BMC Plant Biol., 13*, 158.
[http://dx.doi.org/10.1186/1471-2229-13-158] [PMID: 24119322]

Kaushal, N., Awasthi, R., Gupta, K., Gaur, P., Siddique, K.H.M., Nayyar, H. (2013). Heat-stress-induced reproductive failures in chickpea (*Cicer arietinum*) are associated with impaired sucrose metabolism in leaves and anthers. *Funct. Plant Biol., 40*(12), 1334-1349.
[http://dx.doi.org/10.1071/FP13082] [PMID: 32481199]

Kaushal, N., Bhandari, K., Siddique, K.H.M., Nayyar, H. (2016). Food crops face rising temperatures: An overview of responses, adaptive mechanisms, and approaches to improve heat tolerance. *Cogent Food Agric., 2*, 1-42.

[http://dx.doi.org/10.1080/23311932.2015.1134380]

Kolb, P.F., Robberecht, R. (1996). High temperature and drought stress effects on survival of *Pinus ponderosa* seedlings. *Tree Physiol.,*   *16*(8), 665-672.
[http://dx.doi.org/10.1093/treephys/16.8.665] [PMID: 14871688]

Krishnamurthy, L., Gaur, P.M., Basu, P.S., Chaturvedi, S.K., Tripathi, S., Vadez, V., Rathore, A., Varshney, R.K., Gowda, C.L.L. (2011). Large genetic variation for heat tolerance in the reference collection of chickpea (*Cicer arietinum* L.) germplasm. *Plant Genet. Resour.,*   *9*, 59-69.
[http://dx.doi.org/10.1017/S1479262110000407]

Kumar, J., Kant, R., Kumar, S., Basu, P., Sarker, A., Singh, N. (2016). Heat tolerance in lentil under field conditions. *Legume Genomics and Genetics,*   *7*, 1-11.

Kumar, S., Thakur, P., Kaushal, N., Malik, J.A., Gaur, P.M., Nayyar, H. (2013). Effect of varying high temperatures during reproductive growth on reproductive function, oxidative stress and seed yield in chickpea genotypes differing in heat sensitivity. *Arch. Agron. Soil Sci.,*   *59*, 823-843.
[http://dx.doi.org/10.1080/03650340.2012.683424]

Liu, X., Huang, B. (2000). Heat stress injury in relation to membrane lipid peroxidation in creeping bent grass. *Crop Sci.,*   *40*, 503-510.
[http://dx.doi.org/10.2135/cropsci2000.402503x]

Lopez, C., Soto, M., Restrepo, S., Piégu, B., Cooke, R., Delseny, M., Tohme, J., Verdier, V. (2005). Gene expression profile in response to Xanthomonas axonopodis pv. manihotis infection in cassava using a cDNA microarray. *Plant Mol. Biol.,*   *57*(3), 393-410.
[http://dx.doi.org/10.1007/s11103-004-7819-3] [PMID: 15830129]

Lucas, M.R., Ehlers, J.D., Huynh, B.L., Diop, N.N., Roberts, P.A., Close, T.J. Markers for breeding heat-tolerant cowpea. *Mol. Breed.,*   *31*, 529-536.
[http://dx.doi.org/10.1007/s11032-012-9810-z]

Merga, B., Haji, J. (2019). Economic importance of chickpea: Production, value and world trade. *Cogent Food Agric.,*   *51615718*
[http://dx.doi.org/10.1080/23311932.2019.1615718]

Miller, P., Lanier, W., Brandt, S. (2001). *Using Growing Degree Days to Predict Plant stages. Ag/Extension Communications Coordinator, Communications Services.* Bozeman, MO: Montana State University-Bozeman.

Moy, P., Qutob, D., Chapman, B.P., Atkinson, I., Gijzen, M. (2004). Patterns of gene expression upon infection of soybean plants by *Phytophthora sojae. Mol. Plant Microbe Interact.,*   *17*(10), 1051-1062.
[http://dx.doi.org/10.1094/MPMI.2004.17.10.1051] [PMID: 15497398]

Munier-Jolain, N.G., Ney, B. (1998). Seed growth rate in grain legumes II. Seed growth rate depends on cotyledon cell number. *J. Exp. Bot.,*   *49*, 1971-1976.
[http://dx.doi.org/10.1093/jxb/49.329.1971]

Nadeem, M., Li, J., Wang, M., Shah, L., Lu, S., Wang, X., Ma, C. (2018). Unraveling field crops sensitivity to heat stress: Mechanisms, Approaches and future prospects. *Agronomy (Basel),*   *8*, 128.
[http://dx.doi.org/10.3390/agronomy8070128]

Parameswaran, C., Jagannadham, P.T., Viswanathan, S., Jain, P.K., Ramamurthy, S. (2016). Expression analysis of six chromatin and remodeling complex genes (SWR1) in chickpea in different tissues and during heat stress. *Indian J. Genet. Plant Breed.,*   *76*, 47-56.
[http://dx.doi.org/10.5958/0975-6906.2016.00007.9]

Parankusum, S., Bhatnagar-Mathur, P., Sharma, K.K. (2017). Heat responsive proteome changes reveal molecular mechanisms underlying heat tolerance in chickpea. *Environ. Exp. Bot.,*   *141*, 132-141.
[http://dx.doi.org/10.1016/j.envexpbot.2017.07.007]

Patel, P.N., Hall, A.E. (1990). Genotypic variation and classification of cowpea for reproductive responses to high temperatures under long photoperiods. *Crop Sci.,*   *30*, 614-621.

[http://dx.doi.org/10.2135/cropsci1990.0011183X003000030029x]

Paul, P.J., Samineni, S., Sajja, S.B., Rathore, A., Das, R.R., Chaturvedi, S.K., Lavanya, G.R., Varshney, R.K., Gaur, P.M. (2018). Capturing genetic variability and selection of traits for heat tolerance in a chickpea recombinant inbred line (RIL) population under field conditions. *Euphytica, 214*(2) a.
[http://dx.doi.org/10.1007/s10681-018-2112-8]

Paul, P.J., Samineni, S., Thudi, M., Sajja, S.B., Rathore, A., Das, R.R., Khan, A.W., Chaturvedi, S.K., Lavanya, G.R., Varshney, R.K., Gaur, P.M. (2018). Molecular mapping of QTLs associated with heat tolerance in chickpea. *Int. J. Mol. Sci., 19*(8), E2166 b.
[http://dx.doi.org/10.3390/ijms19082166] [PMID: 30044369]

Porch, T.G. (2006). Application of stress indices for heat tolerance screening of common bean. *J. Agron. Crop Sci., 193*, 390-394.
[http://dx.doi.org/10.1111/j.1439-037X.2006.00229.x]

Prashar, A., Jones, H.G. (2014). Infra-red thermography as a high throughput tool for field phenotyping. *Agronomy (Basel), 4*, 397-417.
[http://dx.doi.org/10.3390/agronomy4030397]

Rani, A., Devi, P., Jha, U.C., Sharma, K.D., Siddique, K.H.M., Nayyar, H. (2020). Developing climate-resilient chickpea involving physiological and molecular approaches with a focus on temperature and drought stresses. *Front. Plant Sci., 10*, 1759.
[http://dx.doi.org/10.3389/fpls.2019.01759] [PMID: 32161601]

Salvucci, M.E., Crafts-Brandner, S.J. (2004). Mechanism for deactivation of rubisco under moderate heat stress. *Physiol. Plant., 122*(4), 513-519.
[http://dx.doi.org/10.1111/j.1399-3054.2004.00419.x]

Shen, C., Que, Z., Xia, Y., Tang, N., Li, D., He, R., Cao, M. (2017). Knock out of the annexin gene OsAnn3 *via* CRISPR/Cas9-mediated genome editing decreased cold tolerance in rice. *J. Plant Biol., 60*(6), 539-547.
[http://dx.doi.org/10.1007/s12374-016-0400-1]

Siddique, K.H.M., Loss, S.P., Regan, K.L., Jettner, R. (1999). Adaptation of cool season grain legumes in Mediterranean-type environments of south-western Australia. *Aust. J. Agric. Res., 50*, 375-387.
[http://dx.doi.org/10.1071/A98096]

Singh, A., Ravikumar, R.L., Jingade, P. (2016). Genetic variability for gametophytic heat tolerance in maize inbred lines. *SABRAO J. Breed. Genet., 48*(1), 41-49.

Singh, N.H., Dhaliwal, G.S. (1972). Effect of soil temperature on seedling emergence in different crops. *Plant Soil, 37*, 441-444.
[http://dx.doi.org/10.1007/BF02139989]

Sultana, R., Choudhary, A.K., Pal, A.K., Saxena, K.B., Prasad, B.D., Singh, S. (2014). Abiotic stresses in major pulses: Current status and strategies. In: Gaur, R.K., Sharma, P., (Eds.), *Approaches to plant stress and their management* (pp. 173-190). New Delhi India: Springer.

Summerfield, R.J., Hadley, P., Roberts, E.H., Minchin, F.R., Rawsthrone, S. (1984). Sensitivity of chickpea to hot temperatures during the reproductive period. *Exp. Agric., 20*, 77-93.
[http://dx.doi.org/10.1017/S0014479700017610]

Thudi, M., Bohra, A., Nayak, S.N., Varghese, N., Shah, T.M., Penmetsa, R.V., Thirunavukkarasu, N., Gudipati, S., Gaur, P.M., Kulwal, P.L., Upadhyaya, H.D., Kavikishor, P.B., Winter, P., Kahl, G., Town, C.D., Kilian, A., Cook, D.R., Varshney, R.K. (2011). Novel SSR markers from BAC-end sequences, DArT arrays and a comprehensive genetic map with 1,291 marker loci for chickpea (Cicer arietinum L.). *PLoS One, 6*(11), e27275.
[http://dx.doi.org/10.1371/journal.pone.0027275] [PMID: 22102885]

Thudi, M., Upadhyaya, H.D., Rathore, A., Gaur, P.M., Krishnamurthy, L., Roorkiwal, M., Nayak, S.N., Chaturvedi, S.K., Basu, P.S., Gangarao, N.V.P.G.R., Fikre, A., Kimurto, P., Sharma, P.C., Sheshashayee, M.S., Tobita, S., Kashiwagi, J., Ito, O., Killian, A., Varshne, R.K. (2017). Genetic Dissection of Drought and

Heat Tolerance in Chickpea Through Genome-wide and Candidate Gene-based Association Mapping Approaches. *PLoS One, 12*(4), e0175609.
[http://dx.doi.org/10.1371/journal.pone.0175609] [PMID: 28384343]

Tsukaguchi, T., Kawamitsu, Y., Takeda, H., Suzuki, K., Egawa, Y. (2003). Water status of flower buds and leaves as affected by high temperature in heat tolerant and heat-sensitive cultivars of snap bean (*Phaseolus vulgaris* L.). *Plant Prod. Sci., 6*, 4-27.
[http://dx.doi.org/10.1626/pps.6.24]

Upadhyaya, H.D., Dronavalli, N., Gowda, C.L.L., Singh, S. (2011). Identification and evaluation of chickpea germplasm for tolerance to heat stress. *Crop Sci., 51*, 2079-2094.
[http://dx.doi.org/10.2135/cropsci2011.01.0018]

Varshney, R.K., Song, C., Saxena, R.K., Azam, S., Yu, S., Sharpe, A.G., Cannon, S., Baek, J., Rosen, B.D., Tar'an, B., Millan, T., Zhang, X., Ramsay, L.D., Iwata, A., Wang, Y., Nelson, W., Farmer, A.D., Gaur, P.M., Soderlund, C., Penmetsa, R.V., Xu, C., Bharti, A.K., He, W., Winter, P., Zhao, S., Hane, J.K., Carrasquilla-Garcia, N., Condie, J.A., Upadhyaya, H.D., Luo, M.C., Thudi, M., Gowda, C.L., Singh, N.P., Lichtenzveig, J., Gali, K.K., Rubio, J., Nadarajan, N., Dolezel, J., Bansal, K.C., Xu, X., Edwards, D., Zhang, G., Kahl, G., Gil, J., Singh, K.B., Datta, S.K., Jackson, S.A., Wang, J., Cook, D.R. (2013). Draft genome sequence of chickpea (Cicer arietinum) provides a resource for trait improvement. *Nat. Biotechnol., 31*(3), 240-246. a
[http://dx.doi.org/10.1038/nbt.2491] [PMID: 23354103]

Vats, S., Kumawat, S., Kumar, V., Patil, G.B., Joshi, T., Sonah, H., Sharma, T.R., Deshmukh, R. (2019). Genome editing in plants: Exploration of technologies advancements and challenges. *Cells, 8*(11), 1-39.
[http://dx.doi.org/10.3390/cells8111386] [PMID: 31689989]

Wahid, A., Gelani, S., Ashraf, M., Foolad, M.R. (2007). Heat tolerance in plants: an overview. *Environ. Exp. Bot., 61*, 199-223.
[http://dx.doi.org/10.1016/j.envexpbot.2007.05.011]

Wang, J., Gan, Y.T., Clarke, F., McDonald, C.L. (2006). Response of chickpea yield to high temperature stress during reproductive development. *Crop Sci., 46*(5), 2171-2178.
[http://dx.doi.org/10.2135/cropsci2006.02.0092]

Waqas, M., Azhar, M.T., Rana, I.A., Azeem, F., Ali, M.A., Nawaz, M.A., Chung, G., Atif, R.M. (2019). Genome-wide identification and expression analyses of WRKY transcription factor family members from chickpea (*Cicer arietinum* L.) reveal their role in abiotic stress-responses. *Genes Genomics, 41*(4), 467-481.
[http://dx.doi.org/10.1007/s13258-018-00780-9] [PMID: 30637579]

Xu, S., Li, J., Zhang, X., Wei, H., Cui, L. (2006). Effects of heat acclimatization pretreatment on changes of membrane lipid peroxidation, antioxidant metabolites, and ultrastructure of chloroplasts in two cool-season turfgrass species under heat stress. *Environ. Exp. Bot., 56*, 274-285.
[http://dx.doi.org/10.1016/j.envexpbot.2005.03.002]

Young, J.C., Zhang, P., Jing, Z., Shi, J. (2003). Genome-wide identification and analysis of heat shock transcription factor family in cucumber (*Cucumis sativus* L). *Plant Omics, 6*, 449.

Zafar, S.A., Hussain, M., Raza, M., Muhu-Din Ahmed, H.G., Rana, I.A., Sadia, B., Atif, R.M. (2016). Genome wide analysis of heat shock transcription factor (HSF) family in chickpea and its comparison with Arabidopsis. *Plant Omics, 9*(2), 136-141.
[http://dx.doi.org/10.21475/poj.160902.p7644x]

Zafar, S.A., Zaidi, S.S., Gaba, Y., Singla-Pareek, S.L., Dhakher, O.P., Li, X., Mansoor, S., Pareek, A. (2019). Engineering abiotic stress tolerance *via* CRISPR/Cas-mediated genome editing. *J. Exp. Bot.,* 1-10.
[http://dx.doi.org/10.1093/jxb/erz476] [PMID: 31644801]

Zhang, Y., Mian, M.A., Chekhovskiy, K., So, S., Kupfer, D., Lai, H., Roe, B.A. (2005). Differential gene expression in *Festuca* under heat stress conditions. *J. Exp. Bot., 56*(413), 897-907.
[http://dx.doi.org/10.1093/jxb/eri082] [PMID: 15710639]

# CHAPTER 5

# Genetic Improvement of Groundnut for Adaptation to Heat Stress Environments

**Murali T. Variath[1], Dnyaneshwar B. Deshmukh[1], Sunil Chaudhari[1], Seltene Abady[1,2], Swathi Gattu[1] and Janila Pasupuleti[1,*]**

[1] *International Crops Research Institute for the Semi-Arid Tropics (ICRISAT), Patancheru, Hyderabad, India*

[2] *Haramaya University, Addis Ababa, Ethiopia*

**Abstract:** Groundnut or peanut (*Arachis hypogaea* L.), an annual legume, is an important oil, food, fodder and feed crop grown in more than 100 countries. Heat and drought stress, and their combination are important abiotic constraints of groundnut production in Asia and Africa, which together accounts over 90% of global groundnut area. An increase in mean air temperature of 2-3 °C is predicted to reduce groundnut yields in India by 23-36% as heat stress during critical stages affects the pod yield. Moreover, heat stress worsens the burden of moisture stress aggravating the pod yield losses. Although groundnut genotypes continue to produce photosynthates under heat stress, only tolerant genotypes possibly have coping mechanisms to partition photosynthates to pods. Understanding the physiological, biochemical, molecular and genetic mechanism of heat-stress tolerance in groundnut is useful to devise breeding strategies to improve adaptation to heat stress. Intense phenotyping of plants grown in the field and glasshouses distinguishes sensitive and tolerant genotypes for heat stress, and to study the associated physiological and morphological differences between such genotypes. This chapter elaborates on the effects of heat stress on different life stages in groundnut, mechanisms contributing to adaptation to heat stress and recent developments in phenotyping, genetics and genomic tools to improve adaptation to heat stress.

**Keywords:** Climate change, Genetics, Groundnut, Heat stress, Mapping, Phenotyping.

## INTRODUCTION

Climate is an important contributing factor to agriculture production and productivity. All living organisms, including microorganisms, plants, animals,

---

[*] **Corresponding author Janila Pasupuleti:** International Crops Research Institute for the Semi-Arid Tropics (ICRISAT), Patancheru, Hyderabad 502324, India. Tel: 9989 9308 55 ; E-mail:p.janila@cgiar.org

**Uday C. Jha, Harsh Nayyar and Sanjeev Gupta (Eds.)**

human beings, *etc.* are exposed to different kinds of stresses. However, plants, due to their sessile nature, are more likely to be affected by abiotic and biotic stresses prevalent in their surrounding environment. Among the abiotic factors, drought, salt, cold, and heat stresses are predominant that individually or in combination can affect crops. Global climate changes have become an important concern in several agro-ecologies necessitating changes in crop and cropping preferences to sustain food security. Depending upon the geographic location, crop yields will be either increased or decreased by climate change. The climate change prediction effects for the semi-arid tropics (SAT) regions indicate a negative trend for yield, thereby threatening food security in these regions (Fischer *et al.*, 2005; Howden *et al.*, 2007). Changes in climatic factors such as temperature, rainfall, day length, light intensity, *etc.* will have a significant influence on plant growth, development and reproductive processes. All plant processes, including germination, seedling emergence, vegetative growth, floral development, pollination, *etc.* have a minimum, maximum and optimum temperature range beyond which these processes are severely impacted. The minimum/maximum temperature below/beyond which irreversible damage can occur is referred to as threshold temperature. If the temperature goes beyond the threshold by even a single degree, then the plant is said to be exposed to heat stress. Reduction in crop yields by 15-35% in Africa and Asia, and by 25-35% in the Middle East is predicted due to increase in temperature by 3-4 °C (Ortiz *et al.*, 2008).

Groundnut (*Arachis hypogaea* L.) or peanut is an economically important oilseed and food crop mainly cultivated by resource-poor small and marginal farmers in the SAT regions of Asia and Africa (Dixon *et al.*, 2001). In most of these regions, the temperatures have already reached the critical limits beyond which the growth processes and productivity of crops are likely to be adversely affected. Groundnut is mainly cultivated in these regions as a rainfed crop with minimal inputs. Nonetheless, rainfall distribution during crop production is critical to ensure higher yields. During the peg development, it is vital that sufficient moisture is present in the soil to allow the pegs to penetrate the soil. The increasing temperatures have already altered the rainfall patterns in many places and this is expected to become more erratic and extreme as the globe warms up (Kumar *et al.*, 2012). The mean temperature in future climates is expected to be 1.5-6.0 °C higher than normal due to global warming, which along with unpredictable day-to-day weather trends would further accentuate the problems of groundnut farmers in these areas (Wheeler *et al.*, 2000; Houghton *et al.*, 2001). The fourth assessment report of the Inter-Governmental Panel on Climate Change (Solomon *et al.*, 2007) has found that the increase in greenhouse gases (GHGs) caused by human anthropogenic activities has resulted in the warming of the climate system by 0.7 °C over the past 100 years; and is projected to rise about 1.8-4.0 °C by 2100. For the South Asia region, the predictions indicate temperature rise of 0.5-

1.2 °C by 2020, 0.9-3.2 °C by 2050 and 1.6-5.4 °C by 2080, depending on future development scenarios.

As in other crops, heat stress can affect all groundnut growth stages, with profound effects on reproduction and seed filling stages (Hamidou *et al.*, 2013), resulting in a significant reduction in yield and/or quality in some growing regions (Akbar *et al.*, 2017). Reproductive stage vulnerability to heat stress can be due to damage to male components, resulting from developmental as well as functional disturbance, such as sucrose and starch accumulation in pollen grains (Sita *et al.*, 2017). Groundnut is geotropic in nature, *i.e.*, flowering, pollination and hybridization occur aerially, but subsequently, the peg containing the developing embryo moves down into the soil and the following processes such as the development of pods and kernels and accumulation of nutrients happen underground. Therefore, both high soil and air temperatures can impact the potential yield as well as kernel quality in susceptible groundnut genotypes. The optimum diurnal temperature requirement for photosynthesis and vegetative growth is between 30 and 35 °C whereas it is much lower (~23 °C) for pod and kernel yield (Cox, 1979). Reduced fruit set and subsequently reduced number of pods and kernel yield were observed when the day temperatures during the reproductive process rose above 35 °C (Ketring, 1984; Prasad *et al.*, 1999a). Studies on the effect of soil temperature on groundnut have revealed that processes such as dry matter accumulation, flower production, the proportion of pegs forming pods, and individual seed mass are affected when the soil temperature exceeds 35 °C (Golombek and Johansen, 1997; Prasad *et al.*, 2000a).

Under such conditions, heat-stress tolerant groundnut genotypes will be needed to counteract the high-temperature stress and sustain productivity in these environments. Screening techniques are available to identify tolerant and susceptible genotypes. Once we identify the stress tolerant genotypes, the next step is to elucidate the mechanism behind heat-stress tolerance. The mechanism by which plants counteract heat stress is a complex process involving many biochemical, molecular and physiological aspects of the plants or their interactions. Depending upon the extremity, duration of stress, plant type, growth stage and other environmental factors in the surroundings, the response mechanism of plants can vary. In this chapter, we have tried to review the effects of heat stress on important life cycle events in groundnut, screening tools that are being deployed to select tolerant genotypes, key traits that are used as a measure of tolerance during screening and elucidate the physiological, biochemical and molecular mechanism behind heat-stress tolerance/avoidance.

# HEAT STRESS AND ITS EFFECT ON DIFFERENT LIFE STAGES OF GROUNDNUT

A crop can be exposed to heat stress at any stage of its growth and development and induce morpho-anatomical, physiological, biochemical and molecular changes, which ultimately lead to poor seed quality and reduced yields. The stress induced can be for short term or long term. Under field conditions, the time of sowing is critical as it will indirectly reflect the stages that are most likely to be exposed to high temperatures. The climatic factors (minimum and maximum temperature, rainfall, average humidity and solar radiation) prevailing during the crop sowing seasons of *Kharif* and *Rabi* at International Crops Research Institute for the Semi-arid tropics (ICRISAT), Patancheru, India is shown in Fig. (**1**). The effects of heat stress on different life stages of groundnut have been widely studied under both controlled and field conditions and some of the common effects are summarized in Table **1** and discussed in this section below.

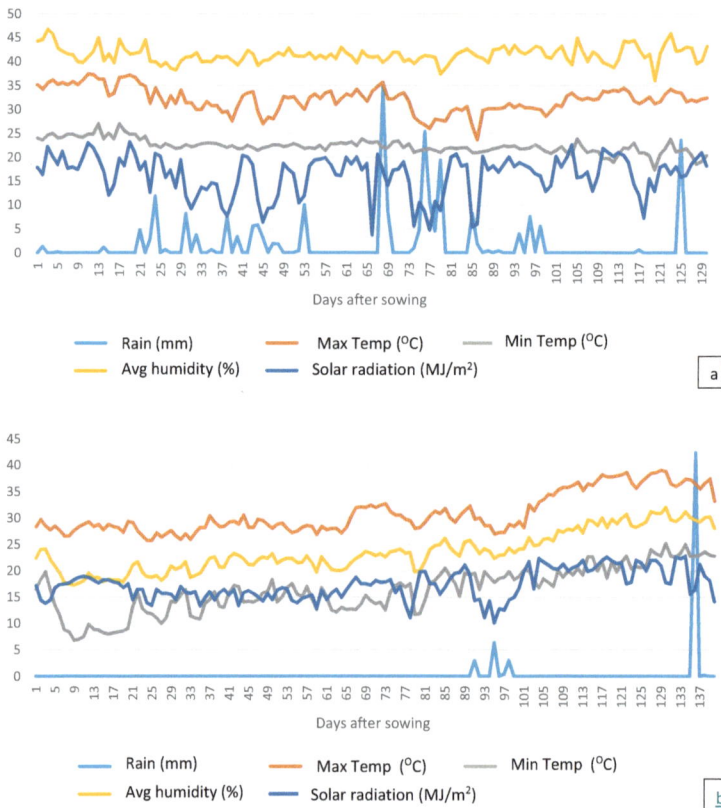

**Fig. (1).** Daily rainfall (mm), maximum temperature, minimum temperature (°C), average humidity (%) and solar radiation (MJ/m²) during **(a)** the post-rainy season of 2013-14 and **(b)** rainy season of 2014.

**Table 1.  Effect of heat stress on different traits studied across life stages of groundnut.**

| Life Stage | Traits Studied | Temperature Treatments | Major Effect of Heat stress | References |
|---|---|---|---|---|
| Seed germination and Seedling stage | Germination percent, seedling vigor, and dormancy | 40, 50, and 60 °C | • Induced seed dormancy by 31% at 40°C<br>• Decreased germination percent and seedling vigor<br>• Heat stress beyond 50°C caused embryonic death | Opio and Photchanachai (2016) |
| | Cell membrane injury, chlorophyll fluorescence, hydroxyproline, galactinol, and unsaturated fatty acids | 40/35 °C (day/night) | • Increased membrane injury, and hydroxyproline<br>• Decreased galactinol, saturated *vs.* unsaturated fatty acids accumulation and chlorophyll fluorescence | Singh *et al.* (2016) |
| | Acquired thermotolerance (ATT) | 50 °C | • Triggered ATT system in the leaf resulting in chlorophyll accumulation upon exposure to light | Selvaraj *et al.* (2011) |
| Vegetative stage | Total leaf area, stem elongation, dry kernel weight, and number of subterranean pegs | 30, 32, and 35 °C | • Decreased individual leaf areas and stem elongation, number of subterranean pegs, and dry kernel weight at 63 and 91 days after planting | Ketring (1984) |
| | Vegetative growth, electrolyte leakage, chlorophyll fluorescence and net photosynthetic rate | 40 to 60 °C | • Increased vegetative growth and electrolyte leakage<br>• Increase photosynthetic rates up to 62% | Talwar *et al.*, (1999b) |
| | Relative injury and chlorophyll fluorescence | 37 and 55 °C | • Increased chlorophyll concentrations after hardening<br>• Reduced relative injury after hardening | Chauhan and Senboku (1997) |
| | Root nodulation and nitrogen fixation | 37 and 40 °C | • Reduced nitrogen fixation by 49%<br>• The function of nodules was adversely affected without a reduction in nodule number | Kishinevsky *et al.*, (1992) |
| | Number of root nodules and nodule mass | | • Reduced nodule number and nodule mass per plant | Lie, (1974) |

| Life Stage | Traits Studied | Temperature Treatments | Major Effect of Heat stress | References |
|---|---|---|---|---|
| Reproductive stage | Days to flowering and flower abortion | 40 to 60 °C | • Diseased days to first flowering by 5-6 days <br> • Increased flower abortion | Talwar *et al.,* (1999a) |
| | Pollen production and viability | 34, 42 and 48 °C | • Reduced pollen viability, pollen number, and pod-set | Prasad *et al.,* (1999a) |
| | Pollen germination and pollen tube growth | 10 to 47·5 °C | • Maximum pollen germination and pollen tube length of genotypes identified as important parameters of tolerant genotypes | Prasad *et al.,* (2001a) Kakani *et al.,* (2002) |
| | Pollen viability, net photosynthesis rate, stomatal conductance, transpiration rate and heat shock proteins (HSPs) | 35/25 and 45/35 °C | • Reduction in pollen viability (up to 45%) and net photosynthesis rate (up to 75%) and stomatal conductance (up to 80%) and Transpiration rate (up to 56%) <br> • Induced production of HSPs | Chakraborty *et al.* (2018) |
| | Crop growth rate (CGR), pod growth rate (PGR), ratio of CGR:PGR, and fruit set | 28 and 38 °C | • Reduced CGR up to 50% <br> • Fruit set reduced from 71 to 58% | Craufurd *et al.,* (2003) |
| | Days to flowering and podding, leaf water potential, hydraulic conductivity, transpiration rate and stomatal conductance | 12, 18, 25, 32 and 40 °C | • Shortened number of days to flowering, podding and maturity <br> • Increased leaf water potential, hydraulic conductivity, transpiration rate and stomatal conductance | Awal *et al.,* (2003) |

*(Table 1) cont.....*

| Life Stage | Traits Studied | Temperature Treatments | Major Effect of Heat stress | References |
|---|---|---|---|---|
| Pod/Seed development and maturity stage | Number of pods per plant, pod yield, and dry matter | 28 and 38 °C | • Reduced dry matter, pod yield and number of pods by 20-28% | Craufurd *et al.*, (2003) |
| | Pod yield and pod growth rate | 35 °C and above | • Reduction pod yield by 1.5- 43.2%<br>• Decreased pod growth rate | Akbar *et al.*, (2017) |
| | Number of pods per plant, pod weight, seed filling | 40 to 60 °C | • Decreased number of pods per plant, seed filling and pod weight | Talwar *et al.*, (1999b) |
| | Number of pods per plant, Kernel Size, days to maturity | 12, 18, 25, 32 and 40 °C | • Reduced number of pods per plant up to 36%<br>• Decreased kernel size<br>• Shortened the time of maturity | Awal *et al.*, (2003) |
| | Pod yield and physiological maturity | 35/25 and 45/35 °C | • Reduced pod yield by 9-35%<br>• Advanced physiological maturity by 15–18 days | Chakraborty *et al.* (2018) |

## Germination, Seedling Emergence, and Vegetative Growth

Germination followed by the emergence and establishment of seedling is the first life-cycle event in groundnut. In the field, the soil temperature is one of the major environmental factors that influence germination, seedling emergence and establishment. In the case of groundnut, additional factors such as seed dormancy, production, harvest and storage conditions are also reported to influence seedling emergence and vigor (Ketring, 1979; Roberts, 1988). The germination event is primarily determined by the temperature and soil moisture in the seeding zone (Kumar *et al.*, 2012). The base temperature, the temperature at which germination is initiated, was found to vary among the genotypes indicating genotypic differences (Wynne and Sullivan, 1978; Mohamed *et al.*, 1988a). The optimum mean soil temperature at which maximum germination and emergence is observed was between 25-30 °C and the process showed a sharp decline beyond 36 °C (Mills, 1964; Mohamed *et al.*, 1988 a,b; Awal and Ikeda, 2002; Prasad *et al.*, 2006; Leong and Ong, 1983). For example, less than 50% emergence of groundnut seedlings was observed at temperatures of 19 and 22 °C, while at temperatures of 25, 28 and 31 °C, the emergence percentage ranged from 70-80% (Leong and Ong, 1983). Soil temperatures beyond 36 °C can induce heat stress and reduce germination and emergence percentage.

Heat-stress influences the vegetative growth of groundnut that includes the production of leaves and an increase in plant biomass. The atmospheric temperature is more important here than the soil temperature. Soil temperature can indirectly affect leaf appearance, branching and flowering as faster seedling emergence occurs at higher soil temperature. Awal and Ikeda (2002) reported that the leaf development in groundnut plants grown in warmer soils was faster than those grown under cooler conditions. With rising air thermal times (134-1147 °C d), the number of main stem leaves increased but decreased the ratio of the number of main stem leaves to the total number of leaves.

Specific leaf area (SLA) is an important surrogate trait that can affect canopy expansion and growth under heat stress. At high temperatures, the total leaf area per plant is reduced, which affects light interception capacity and light use efficiency of susceptible genotypes (Kumar *et al.*, 2012). Day/night temperatures of 35/22 °C decreased individual leaf areas and dry weights at 63 and 91 days after planting in two groundnut cultivars- Tamnut 74 and Starr with about 49% reduction in leaf area per plant for Tamnut 74 and about 80% for Starr (Ketring, 1984). On the other hand, Talwar *et al.* (1999b) observed greater vegetative growth among three groundnut genotypes (ICG 1236, ICGS 44 and Chico) exposed to 35/30 °C day/night temperatures. Greater vegetative growth expressed as the crop growth rate (CGR) was observed among groundnut genotypes exposed to 35 °C air temperatures under field conditions (Akbar *et al.*, 2017). The CGR was found to be reduced in heat susceptible genotypes at 38 °C soil temperature, whereas no effect was observed in the tolerant genotypes ICGV 86021, 796 and ICG 1236. Thus, CGR can be used as a parameter for selecting genotypes tolerant to high soil temperature stress (Craufurd *et al.*, 2003).

**Reproductive Growth**

As compared to other life stages, the reproductive processes in groundnut, such as pollen germination, growth of pollen tubes, and the number of pollen grains retained by the stigma, are extremely sensitive to heat stress and are affected at temperatures >28 °C (Prasad *et al.*, 2001a). For percentage pollen germination, the mean minimum, optimum and maximum cardinal temperatures ($T_{min}$, $T_{opt}$ and $T_{max}$) were 14.1, 30.1 and 43.0 °C whereas, for maximum pollen tube length, the parameters were 14.6, 34.4 and 43.4 °C, respectively (Oakes, 1958; Prasad *et al.*, 2000a; Kakani *et al.*, 2002). Exposure to very high temperatures of ≥39 °C resulted in 30-100% reduction in the number of flowers that could form pegs and set pods compared to 28 °C in controls (Ketring, 1984; Prasad *et al.*, 1999 a, b, 2000a, 2001b). For example, exposure to high temperatures of 33 and 40 °C for 12 h resulted in only 60 and 10% of flowers setting fruits in the cultivar ICGV

86015, attributed to the presence of fewer pollen grains and poorer pollen viability (Prasad *et al.* 1999a, 2001a).

## Nodulation and Nitrogen Fixation

Groundnut being a leguminous crop largely meets its nitrogen requirement at different growth stages through symbiotic nitrogen fixation. Heat and moisture stress can adversely affect the *rhizobium*-legume symbiosis. High soil temperatures have a significant effect on growth and survival of rhizobia, formation of root hairs, formation of infection threads, structure and development of root nodules, and activity of the nitrogenase enzyme compared to high air temperature. The influential consequences are reduction in nodule numbers, nodule mass per plant and total dry matter production (Lie, 1974). In most of the tropical and subtropical regions of the world where groundnut is predominantly grown, the soil and air temperatures are often between 35 to 40 °C, resulting in reduced root hairs and biomass and production of thin and unbranched roots. Hence, groundnut plants susceptible to heat stress produce fewer nodules per plant. Soil temperature beyond 40 °C completely inhibits the *Bradyrhizobium* symbiosis through blocking the process of nodulation and nodules function (Kishinevsky *et al.*, 1992). The total nitrogen fixation was reduced by 49% without a reduction in nodulation or nodule number under continuous exposure of 37 °C soil temperature due to a substantial reduction in the activity of nitrogenase enzyme in the nodules (Prasad *et al.*, 2001b). The optimum root temperature for active nitrogen fixation, high plant dry weight and total nitrogen is 30°C (Prasad *et al.*, 2001b). Apart, heat stress, moisture availability during the crop growth period, especially the first 50 days after emergence, drastically reduces root growth. High soil temperature, coupled with moisture stress, can kill most of the rhizobia and free-living nitrogen-fixing bacteria (Kishinevsky *et al.*, 1992). Water stress reduces nitrogen fixation through a reduction in the activity of nitrate reductase enzyme. Thus, genotypes with drought tolerance often possess larger root density and volume, and goes deeper when grown under mild stress conditions could be efficient in nitrogen fixation.

## Peg, Pod and Seed Development

The peg formation and number of pegs are sensitive to high day/night temperatures. Reduction in peg numbers ranged from 5.0 to 7.7 per plant at air temperatures above 28 °C during the day and 22-28 °C during the night (Prasad *et al.*, 1999b). In contrast, the soil temperature did not affect pegs number but had a significant influence on the formation and number of pods and seeds. The overall accumulation of dry matter or biomass is limited by heat stress thus reducing the pod number and development. Pod growth rate (PGR) used as a measure of pod development was found to decrease in susceptible genotypes under high soil and air temperatures (Craufurd *et al.* 2003; Akbar *et al.* 2017). The geocarposphere temperature between 21 and 29 °C is optimum for higher yield and quality of the produce. An increase in canopy temperature by 6 to 9 °C above the optimum temperature (28 °C) and/or by 3 to 4 °C in the podding zone above 23 °C during the reproductive stage adversely affects pod development and maturity leading to reduced harvest index (Craufurd *et al.*, 2002). Pod and seed dry weight is reported to be reduced at soil temperatures >33 °C in the podding zone. The pod yield was significantly reduced when the plants were exposed to 38/32 °C (day/night) soil temperature compared to 26/20 °C and 32/26 °C due to adverse effects on the rate of pod initiation and growth, and development of sound mature kernels. Apart from these, heat stress also affects the nutritional composition leading to reduced nutritional value to the consumers, and reduced germination percent and seedling vigor. The findings of some controlled environments studies revealed that the high soil temperature (>37°C) at the podding stage significantly reduces the number of mature pods, kernel size, hundred kernel weight resulting in reduced shelling outturn and kernel yields (Dreyer *et al.*, 1981; Golombek and Johansen, 1997; Prasad *et al.*, 2000b).

## Yield and Associated Traits

The effects of heat stress on yield losses in legumes, cereals and other crops, including groundnut is well-documented (*e.g.*, Paulsen, 1994; Prasad *et al.*, 2000b; Hall, 2004; Akbar *et al.*, 2017). The time of imposition and duration of heat stress can have a significant effect on yield and its associated traits. Optimum air and soil temperatures at flowering, podding and seed development stage are critical to maintain high yields. Heat stress mainly impacts the yield through decreased fruit-set, early-maturity or decreased shelling outturn through the poor pod and seed development. A decrease in fruit-set can occur if the crop is exposed to high day temperatures of >35 °C at the reproductive stage, mainly due to poor pollen viability, reduced pollen production and poor pollen tube growth resulting in poor fertilization of flowers (Prasad *et al.*, 1999b, 2000a, 2001a). This reduction in pod yield can range from 1.5% to 43.2% in susceptible groundnut genotypes whereas, tolerant groundnut genotypes recorded 9.0% to 47.0% increased pod yield along with 0.65% to 3.6% increase in pod growth rate (Akbar *et al.*, 2017).

Daily temperatures of >38 °C at the pod/seed development stage can influence a susceptible genotype to mature early by accelerating the seed filling stage which negatively impacts the yield potential (Boote *et al.*, 2005). Groundnut cultivars

ICGV 86105 (Spanish Bunch) and ICGV 87282 (Virginia Bunch) exposed to high air day/night (38/22 °C) and/or high soil day/night (38/30 °C) temperatures at flowering and podding stage caused a significant reduction in dry matter production, partitioning of dry matter to pods, and pod yields in both the genotypes. The Virginia cultivar was more sensitive to heat stress with pod yield reductions of 49-59% in comparison to 21-24% for the Spanish cultivar (Prasad *et al.*, 2000b).

For yield associated traits such as seed yield and shelling outturn there was a decrease of 334 gm$^{-2}$ to 30 gm$^{-2}$, and 82% to 74% respectively, as the day time maximum/night time minimum temperatures were increased from 32/22 to 36/26, 40/30 and 44/34 °C (Prasad *et al.*, 2003). In the case of seed number per pod and seed size, the decrease was by 0.05 seeds per pod and 0.0075 g, respectively (Prasad *et al.*, 2003). Hence, 32/22 °C is optimum in groundnut to get a good yield and beyond this there will be yield penalties in susceptible genotypes. Similarly, Talwar *et al.* (1999a) recorded 41-62% shelling outturn in three groundnut genotypes exposed to day/night temperatures of 35/30 °C in comparison with 60-76% at 25/25 °C. One of the possible explanations for this reduction could be the reduced partitioning of dry matter into seeds due to heat stress, as stated in some studies (Crauford *et al.*, 2002; Prasad *et al.*, 2000b; Akbar *et al.*, 2017). High soil temperatures reduce kernel size leading to reduced shelling outturn when stress is imposed at the podding stage (Ketring, 1984; Talwar *et al.*, 1999a, Prasad *et al.*, 2000b).

## MECHANISM OF HEAT-STRESS TOLERANCE

To combat the effects of heat stress, plants have evolved tolerance and/or avoidance mechanisms depending on the intensity, duration and rate of temperature change to which they are frequently exposed (Wahid, 2007). In response to short term stress, plants have developed avoidance or acclimation mechanisms that include physical changes within the plant architecture such as modifying leaf orientation, transcriptional cooling, altering membrane lipid composition, reflecting solar radiation, extensive rooting, leaf shading of tissues that are sensitive to sunburn and often altering metabolic activity. Heat acclimatization can also be induced in the plants by exposure to moderately elevated, non-lethal temperature stress for a specific time duration (Selvaraj *et al.*, 2011; Baurle, 2016). Heat-stress tolerant plants have developed novel mechanisms which include physiological, biochemical and molecular alterations in plant activity, to counteract the negative effects arising from exposure to heat for long durations. This includes altering of metabolic activity through the production of compatible solutes, maintaining cell turgor through osmotic adjustments and modifying the antioxidant system. At the molecular level, plant induces the expression of genes that function as osmoprotectants, detoxifying enzymes, transporters, and regulatory proteins (Semenov and Halford, 2009; Krasensky and Jonak, 2012) that enables plants to tolerate heat stress. Besides these innate plant mechanisms, exogenous application of osmoprotectants (proline; glycine betaine; trehalose, *etc.*), phytohormones (abscisic acid, gibberellic acids, jasmonic acids, brassinosteroids, salicylic acid, *etc.*), signaling molecules (nitric oxide), polyamines (putrescine, spermidine, and spermine), trace elements (selenium, silicon, *etc.*) and nutrients (nitrogen, phosphorus, potassium, calcium, *etc.*) have been found effective in mitigating heat-stress induced damage in plants (Hasanuzzaman *et al.*, 2010, 2012, 2013 a,b; Waraich *et al.*, 2012).

The information on physiological, biochemical and molecular mechanisms involved in contributing heat-stress tolerance in groundnut is limited. In the preceding sections, we discuss a few of these mechanisms that are operating in groundnut.

## Short-term Avoidance/Acclimatization Mechanisms

Studies on the response of wild *Arachis* accessions to heat stress have revealed that leaf hairiness and thickness, modification in membrane lipid composition and regulation of transpiration rate along with epicuticular wax load are some of the factors that help plant to avoid/acclimatize under heat stress (Nautiyal *et al.*, 2008; Saffan, 2008). Heat-induced injuries were lower in genotypes with low SLA indicating that genotypes with thick leaves are more tolerant compared to thin leaves. Genotypes with higher wax load in their leaves have increased reflectance capacities which enables them to avoid the extra heat (Nautiyal *et al.*, 2008). When groundnut seedlings were exposed to different temperatures (25, 35 and 40 °C), seedlings exposed to 40 °C were found to contain a higher proportion of polyunsaturated fatty acids than saturated (Saffan, 2008). Apart, other biochemical constituents such as iodine, peroxide value, phenolic content, tocopherol, β-sitosterol, campesterol and stigmasterol content also increased at 40 °C temperature.

## Tolerance Mechanisms

High-temperature stress (>38.0°C) triggers the evaporation of available moisture in the field leading to water deficit stress for crop plants. During the daytime, high temperature leads to a high rate of transpiration in plants resulting in lower water potential and disturbances in normal metabolic processes (Tsukaguchi *et al.*, 2003). The physiological ability of a plant to tolerate heat stress is directly correlated with their ability to maintain net photosynthetic rate and cellular membrane integrity, accumulate compatible osmolytes, and efficiently partition the nutrients among the different plant components (Nagarajan *et al.*, 2010; Scafaro *et al.*, 2010; Akbar *et al.*, 2017). Genetic variability for membrane injury, leaf biochemical constituents, epicuticular wax load, photosynthesis rate, production of reactive oxygen species (ROS) and heat shock proteins (HSPs) have been reported among cultivated and wild groundnut species under heat stress (Nautiyal *et al.*, 2008; Singh *et al.*, 2016). The tolerance mechanisms reported in plants are discussed with reference to groundnut.

## Cell Membrane Injury (CMI)

The integrity and functions of biological membranes are affected in response to heat stress. Plasma membranes, an important component of plant cells are reported as highly sensitive to heat stress and are often primary sites for injury in many crop plants. However, photosynthetic apparatus is reported to be more sensitive to heat stress than plasmalemma integrity in groundnut (Talwar *et al.*, 1999b). Heat stress can alter the tertiary and quaternary structures of membrane proteins or increase the proportion of unsaturated fatty acids resulting in increased fluidity of the membranes (Savchenko *et al.*, 2002). Such alterations of membrane structure lead to activation of lipid-based signaling cascades, increased $Ca^{2+}$ influx, and cytoskeletal re-organization (Bita and Gerats, 2013). This can cause organic and inorganic ions leakage from affected cells due to cell membrane injury and in severe cases even cell death. Cell membrane thermostability (CMT) is measured using *in vitro* leaf disc method (Ketring, 1985), or by cell suspension culture (Venkateswarlu and Ramesh, 1993) which tests membrane thermostability at elevated temperatures by measuring the electrolyte leakage from damaged leaf cells. The increased solute leakage is an indication of decreased CMT and is used as an indirect measure of heat-stress tolerance in many crop plants including groundnut. CMT is often influenced by the age of the plant/tissue, sampling organ, stage of development, growing season and degree of hardening (Kaushal *et al.*, 2016).

In groundnut, around 18% increased CMI was reported when seedlings exposed to day/night temperatures of 40/35 °C (Singh *et al.*, 2016). Talwar *et al.* (1999a) reported that high temperature acclimatized groundnut plants (grown under 35/30 °C) had higher membrane thermostability and decreased thermoinhibition than the control. Genetic variability was reported for CMI under heat stress among wild accessions of groundnut (Nautiyal *et al.*, 2008). The CMI is reported to be increased among all accessions with increase in temporal heat-intensity. The wild accessions *A. glabrata* 11824 had the lowest injury (42%) compared to *A.*

*appresipilla* 11786 (85%) (Nautiyal *et al.*, 2008). The ability of the groundnut cultivar ICGS 44 to better maintain CMT under heat stress could be attributed to its lesser sensitivity to heat stress as compared to susceptible genotypes DRG 1 and AK 159 (Chakraborty *et al.*, 2018).

### Accumulation of Compatible Osmolytes under Heat Stress

Plants accumulate certain low molecular weight organic compounds called thermoprotectants such as sugars (trehalose), polyols, proline and glycine betaine in response to heat stress (Hare *et al.*, 1998; Sakamoto and Murata, 2002). These thermoprotectants are reported to have protective functions in heat-stressed cells (Jain *et al.*, 2007; Rasheed *et al.*, 2011), and contribute as one of the important mechanisms to acquire thermo-tolerance in plants (Rasheed *et al.*, 2011). Glycine betaine, an important thermoprotectant plays a crucial role under heat or salinity stress (Sakamoto and Murata, 2002). Apart, proline has also been widely reported as an osmoprotectant in response to abiotic stresses in many crop plants (Kishore *et al.*, 2005). Glycine betaine and proline helps in buffering the cellular redox potential under heat stress (Wahid and Close, 2007). Transient elevation of free $Ca^{2+}$ in the cytoplast acts as a regulator of many physiological and biochemical processes in response to heat stress in groundnut (Yang *et al.*, 2013). The details on role of osmolytes such as HSPs, proline, glycine betaine, trehalose, brassinosteroids, *etc.* in offering thermo-tolerance are available in Kaushal *et al.* (2016).

In groundnut, maintenance of high relative water content, and accumulation of soluble sugars, free amino acids, and proline under water stress protect the cell by balancing the osmotic strength of cytosol with the help of vacuole reservoirs (Padmavathi and Rao, 2013). However, under heat stress, a significant increase in inositol and hexoses (glucose and fructose) content and reduction (>50%) in sucrose content in leaf tissues indicates the degradation of storage carbohydrates for improved osmotic adjustment in tolerant groundnut genotypes (Chakraborty *et al.*, 2018). Metabolite profiling of Virginia-type groundnuts grown under 40/35°C have revealed that the combination of several metabolites explained 41 to 61% of the differential physiological response in heat-stressed groundnut seedlings (Singh *et al.*, 2016). Genetic variability for concentration of compatible solutes such as sugars, proline and total amino acid was reported among heat tolerant and susceptible wild *Arachis* accessions (Nautiyal *et al.*, 2008).

### Photosynthetic Characteristics under Heat Stress

Photosynthesis is one of the key plant functions to be affected by temperature changes. Photosynthesis can withstand temperatures of 30-35 °C but is adversely affected by damages to chloroplast stroma and thylakoid membrane above 40 °C (Schuster and Monson, 1990; Wang *et al.*, 2010). The net photosynthetic rate, stomatal conductance, transpiration rate and $F_v/F_m$ (ratio of variable to maximum fluorescence) was significantly reduced in groundnut genotypes exposed to 36.3/19.5 °C and 40.4/23.0 °C day/night temperature regimes (Chakraborty *et al.*, 2018). They reported that tolerant genotypes (*e.g.*, ICGS 44) maintained the highest leaf gas exchange capacity and efficiency of photosystem II at temperatures beyond 40 °C due to which it could significantly cool down its canopy and maintain an optimum metabolic conditions in leaf tissues (Chakraborty *et al.*, 2018). The ability of heat-stress tolerant genotypes to sustain leaf gas exchange characteristics and maintain greater vegetative growth and photosynthetic rates under heat stress was also reported by Talwar *et al.*, (1999b) and Nagarajan *et al.*, (2010) indicating that these traits could be used as heat-stress tolerance surrogate traits for selection decisions in the breeding program. Similarly, significant variability among the genotypes reported for chlorophyll a ($Chl_a$), $Chl_b$ and $Chl_{(a+b)}$ concentrations, and $Chl_{(a/b)}$ ratio under heat stress indicating that it can also be used as selection criteria to identify heat-stress tolerant genotypes after further validations (Chauhan and Senboku, 1997). The role of maintaining cooler canopy temperatures in heat-stress tolerance is also documented in other crop species (Rebetzke *et al.*, 2013; Fu *et al.*, 2016). Acclimatized groundnut plants showed increased photosynthetic rate and lower reduction of $F_v/F_m$ ratio when exposed to 50 or 55 °C temperature indicating their better heat-stress tolerance capacity compared to non-acclimatized plants which showed complete thermoinhibition (Talwar *et al.*, 1999b).

## *Assimilate Partitioning Inside Plant System under Heat Stress*

Heat stress decreases the grain yield in plants by hampering the source-sink movement which takes place *via* symplastic and apoplastic pathways. These pathways play significant roles in the partitioning of assimilates under heat stress (Taiz and Zeiger, 2006). Although groundnut accumulates higher photosynthates in response to heat-stress (~40°C), the partitioning of photosynthates is severely affected under high temperature (Akbar *et al.*, 2017). Increase in temperature regime during the crop growth period results in early maturity as the plants tends to complete their entire life faster under stress. This could be attributed to temperature-induced restriction in metabolic activities which also causes 1.5 to 43.2% yield losses in groundnut under heat-stress environments (Chakraborty *et al.*, 2018). The heat-tolerant genotypes showed a marginal reduction in the pod growth rate under heat stress without much difference in phenophases and yield. Hence, pod yield and/or yield penalty due to heat stress can be used as a criterion to select genotypes with improved tolerance and higher yields under heat stress (Akbar *et al.*, 2017).

## Reactive Oxygen Species (ROS) and Antioxidant Potential Under Heat Stress

Production of ROS in response to abiotic stress such as heat, drought, salinity, *etc.* constitutes a major component of plant response. Under heat stress conditions, various ROS like singlet oxygen ($^1O_2$), superoxide radical ($O_2^-$), hydroxyl radical ($OH^-$), and hydrogen peroxide ($H_2O_2$) are produced in the plant system which causes oxidative damage in the cells (Liu and Huang, 2000). The high concentrations of ROS can cause lipid peroxidation in cell membranes measured in terms of malondialdehyde or 2-thiobarbituric acid reactive substances content (Xu *et al.*, 2006). Other potential effects of increased ROS could be protein oxidation, DNA damage, and destruction of pigments (Hasanuzzaman *et al.*, 2012; Xu *et al.*, 2006). The production of ROS under heat stress occurs in photosystem I and II of chloroplasts, plasma membrane, mitochondria, peroxisomes, apoplasts, and endoplasts (Soliman *et al.*, 2011). To combat with increased ROS production, tolerant plants induces an antioxidant system through the production of enzymes such as superoxide dismutase, catalase, peroxidase, glutathione reductase and ascorbate peroxidase, and other metabolites such as reduced glutathione, ascorbic acid, tocopherols, and carotenoids components as ROS scavenging agents (Kaushal *et al.*, 2016). Superoxide dismutase mediated detoxification of superoxide radicals results in the synthesis of hydrogen peroxide which is further scavenged by catalase and ascorbate peroxidase.

Information on the effects of heat stress on ROS production in the groundnut is lacking. There is an urgent need to characterize genotypes based on ROS production under heat stress conditions, as it is the first line of response to any abiotic stress in many crops. Studies on the response of groundnut cultivars to water deficit stress have revealed that production of $H_2O_2$, superoxide radical content and lipid peroxidation increased in susceptible cultivars DRG 1, AK 159 and ICGV 86031 and decreased in tolerant cultivars ICGS 44 and TAG 24 (Chakraborty *et al.*, 2015). A negative correlation was observed between the production of ROS and the activities of different antioxidant enzymes such as superoxide dismutase, catalase, peroxidase, ascorbate peroxidase and glutathione reductase under water deficit stress (Chakraborty *et al.*, 2015). As water deficit stress is one of the indirect effects of heat stress, a similar response may exist for heat stress in terms of ROS accumulation and antioxidant defense as reported. However, further research regarding the involvement of signaling molecules for improving the antioxidant system under heat stress is required in groundnut.

## *Production of Heat Shock Proteins (HSP)*

Production of HSPs is important to protect plants from the deleterious effects of heat-stress. HSPs are known to exist in all groups of living organisms and mainly involved in the functional stability of existing proteins and preventing denaturation of newly synthesized proteins under heat-stress (Wang *et al.*, 2004; Tripps *et al.*, 2009; Usman *et al.*, 2014). These are classified as large (60–110 kDa) and small HSPs (15–45 kDa) based on their molecular weight (Young, 2010). The larger HSPs families such as HSP 70 and HSP 90 function as molecular chaperone by upregulating several downstream genes associated with heat tolerance in plants (Kumar *et al.*, 2016).

The rapid induction of small (HSP 17 and HSP 40) and large molecular weight (HSP 70 and HSP 90) HSPs correlated with long-term heat-stress tolerance is reported in groundnut (Chakraborty *et al.*, 2018). The study further revealed that the tolerant and moderately tolerant genotypes ICGS 44 and GG 7, respectively had higher induction of HSPs for relatively longer time duration compared to susceptible genotypes DRG 1 and AK 159 (Chakraborty *et al.*, 2018).

## SCREENING TECHNIQUES FOR HEAT-STRESS TOLERANCE

An effective screening technique that helps to distinguish susceptible from tolerant genotypes is one of the prime requirements for heat-stress tolerance breeding and/or mapping studies. One of the major reasons for low success rate in breeding for heat-stress tolerance in groundnut is the lack of an efficient high throughput screening technique useful to select the parents for crossing nurseries and superior recombinants from segregating populations. The screening methods that are currently being used to identify heat-stress tolerant genotypes are discussed below.

### Field Screening under Heat-Stress Environment

This is the most widely followed method for identifying heat-stress tolerant genotypes. Here, the genotypes are evaluated under high temperature conditions directly in the field by altering the sowing dates and comparing them with the normal sowing season. An advantage of this method is that the genotypes identified by this study is expected to perform well under heat stress but cannot be taken as proof of cause-and-effect due to the dynamic nature of temperature which varies from season-to-season and fluctuates on a daily basis (Dreccer *et al.*, 2015).

## Glasshouse Screening Method

The response to heat stress is highly variable across developmental stages and physiological processes (Driedonks *et al.*, 2016). In this regard, greenhouse screening is an effective method as it offers an opportunity to evaluate plants throughout their life cycle (Janila and Nigam, 2013). Different environmental factors including temperature can be controlled under glasshouse and is useful to study the plants response to different duration and intensity of heat stress across life stages. It is also possible to precisely elucidate the cause of heat-stress tolerance in a particular genotype, as all the other environmental factors are controlled.

## Temperature Induction Response

Germinated seedlings are exposed to sublethal temperatures (38-54 $^{0}$C) for 5 hours and at lethal temperatures (58 $^{0}$C) for 3 hours to screen genotypes for heat-stress tolerance (Rani *et al.*, 2018). The identified thermotolerant seedlings are transferred to the field and their progenies are evaluated using recurrent selection. An advantage of the technique is the ability to screen large populations. However, there is a possibility of chance escapes as plants in some cases can develop adaptation mechanism which enables them to cope with short-term heat stress for a short period. For better results, the identified thermotolerant seedlings need to be evaluated in field conditions under situations of high temperature.

## Detached Leaf Assay

The fully expanded leaf from the third/fourth node is detached from the plant and this leaf is then exposed to high-temperature stress. It is a quick and effective method and offers rapid and accurate screening of a large number of genotypes in a relatively shorter time (12 days per cycle). It is also useful for phenotyping of the mapping population to identify quantitative trait loci (QTLs) controlling acquired thermotolerance (Selvaraj *et al.*, 2011).

## HEAT-STRESS TOLERANT GENOTYPES OF GROUNDNUT

Heat-stress tolerance is a complex trait with polygenic inheritance and is highly influenced by environmental variations (Sadat *et al.*, 2013). Identification of tolerant genotypes, surrogate traits, and their genetic nature are of paramount importance to devise breeding strategies. Substantial genetic variation exists for heat-stress tolerance in groundnut through partitioning of dry matter to pods and kernels (Craufurd *et al.*, 2002), fruit set (Prasad *et al.*, 2001a), membrane thermostability (Srinivasan *et al.*, 1996), chlorophyll fluorescence (Chauhan and Senboku, 1997) and acquired thermostability (Gangappa *et al.*, 2006; Selvaraj *et al.*, 2011). Exploration and utilization of genetic variations present in landraces, obsolete and cultivated varieties, breeding lines and wild accessions are vital for improvement of heat-stress tolerance in groundnut. Some heat-stress tolerant groundnut genotypes along with surrogate traits are summarized in Table **2**.

Table 2.  **Heat-stress tolerance sources and associated traits/indices in groundnut.**

| Heat Tolerance Sources | Heat Tolerance Associated Traits/Indices | Reference |
|---|---|---|
| 55-437, 796, GJG 31, ICGV 87846, ICGV 03057, ICGV 07038 and GG 20 | Pod yield, hundred seed weight, pod growth rate and partitioning | Ntare *et al.*, (2001); Akbar *et al.*, (2017) |
| ICGS-76, COCo38 and COCo68, ICGV 86707 | Relative injury | Selvaraj *et al.*, (2011); Chauhan and Senboku, (1997) |
| 796, 55-437, ICG 1236, ICGV 86021, ICGV 87281 and ICGV 92121 | Crop growth rate, pod growth rate and partitioning | Craufurd *et al.*, (2002) |
| K 134, TMV 2, Dh 40, GN2, GN4, GN5, GN11 | Seedling thermotolerance/survival | Naidu *et al.*, (2017); Rani *et al.*, (2018) |
| TIR-21, TIR-34, JAL-07 and CSMG 84-1 | Cell membrane thermostability and chlorophyll fluorescence | Babitha *et al.*, (2006) |
| 55-437, ICGV 93232, ICGV93233, ICGV 93255, TMV 2, ICG 1236,and ICGV 86015 | Pollen germination and pollen tube growth | Kakani *et al.*, (2002) |
| SPT 06-07 | Membrane injury, chlorophyll fluorescence and metabolites (including hydroxyproline, galactinol, saturated and unsaturated fatty acid ratio) | Singh *et al.*, (2016) |

## KEY TRAITS FOR SELECTING HEAT-STRESS TOLERANT GENOTYPES

Identification of key traits that contribute to stress tolerance is critical for the success of a breeding program. Among different traits, pod yield under stress is the best indicator of a plant's ability to respond to adverse conditions. A genotype with minimum yield penalty under stress and non-stress condition is considered promising. However, in groundnut the pods remain inside the soil and it is very difficult to predict the yield performance of a genotype until after harvest. Hence, surrogate traits that contribute to better yield performance under stress in the initial stages needs to be identified and used in the breeding program. This will enable the selection and advancement of genotypes/populations in the early stages and hence, economic utilization of resources and time. Different traits are used to screen genotypes for stress tolerance. Some commonly used traits is discussed below.

## Pod Yield and Associated Traits

Stable pod yield performance under heat stress ($\geq 35^0C$) has been the most widely used selection criteria to identify tolerant genotypes in groundnut (Akbar *et al.*, 2017; Ni *et al.*, 2018). A positive correlation exists among pod, kernel and haulm yields, crop growth rate (C), pod growth rate (R), partitioning coefficient (*p*) and harvest index in groundnut (Frimpong, 2004). Genotypes grown under heat stress recorded greater vegetative growth and higher photosynthetic rates leading to more accumulation of photosynthates (Talwar *et al.*, 1999b). However, only the heat tolerant genotypes are capable of partitioning the photosynthates to pods (Craufurd *et al.*, 2002; Kakani *et al.*, 2015). Akbar *et al.* (2017) evaluated 62 groundnut genotypes under stress and non-stress conditions in the field and identified ICGVs 07246, 03042, 06039, 07012, 06040, 06424, and 07038 as heat-tolerant based on stress tolerance index and pod yield performance. The crop growth rate (C), pod growth rate (R), and partitioning coefficient (p) can be estimated following a modified procedure suggested by Williams and Saxena (1991):

$$C = (HWT + (PWT \times 1.65) / T_2),$$

$$R = (PWT \times 1.65) / (T_2 - T_1 - 15),$$

$$p = R/C$$

Where *HWT* is the haulm weight, *PWT* is the pod weight, $T_1$ is the number of days from sowing to harvest, $T_2$ is the number of days from sowing to flowering, and 15 is a constant reflecting the number of days between flowering and the start of pod expansion (Ntrare *et al.*, 2001). The harvest index can be estimated as the ratio of PWT and (PWT +HWT) (Frimpong, 2004).

## Heat Susceptibility Index

Heat susceptibility index (HSI) is another important parameter used to screen genotypes under field conditions. HSI estimates the total effect of heat in terms of reduction in pod yield and other yield attributing quantitative traits. It can be calculated according to the formula given by Fisher and Maurer (1978).

HSI= [1-YD/YP]/D

Where,

YD=Mean pod yield of the genotype in stress environment, YP=Mean pod yield of the genotype under non-stress environment and D is stress intensity which is calculated as follows:

D = 1-[Mean YD of all genotypes/Mean YP of all genotypes]

## Cellular Membrane Thermostability

Cellular membrane thermostability (CMT) is also used to screen genotypes for heat-tolerance in groundnut. The genetic variability for this trait in groundnut was reported by Craufurd *et al.* (2003). The CMT can be measured using the procedure described by Martineau *et al.* (1979). The CMT test is based on the fact that the injury imposed on leaf tissues under heat-stress weakens the cell membrane leading to leakage of electrolytes out of the cell (Nyarko *et al.*, 2008). The cell membrane thermostability is measured using the degree of injury which is induced due to heat-stress *i.e.* relative injury (RI) can be calculated as follows:

Relative Injury [RI] (%) = $\{1 - [1 - (H_1/H_2)]/ [1 - (C_1/C_2)]\} \times 100$

Where H and C refer to the conductance in heat treatment and control tubes, and subscripts 1 and 2 refer to readings before and after autoclaving, respectively. The conductance in treatment tubes (*i.e.* $H_1/H_2$) is a measure of electrolytes leaked from cells due to high temperature and is assumed to be proportional to the degree of injury to membranes (Martineau *et al.*, 1979).

## Chlorophyll Fluorescence

Chlorophyll fluorescence, which is measured as a ratio of variable ($F_v$) to maximum ($F_m$) chlorophyll, is another trait often used to screen genotypes for heat-stress tolerance (Singh *et al.*, 2016). Electrolyte leakage, variable chlorophyll fluorescence or photosynthesis impairment and pollen fertility are found to be related with pod yield (Devasirvatham *et al.*, 2016). Of these, electrolyte leakage and chlorophyll fluorescence ratio are negatively correlated (Srinivasan *et al.*, 1996) whereas, chlorophyll fluorescence and chlorophyll concentration are positively correlated in groundnut (Chauhan and Senboku, 1997).

## Pollen Characteristics

Pollen fertility, germination and pollen tube growth are sensitive to high temperature (39 °C) in groundnut (Prasad *et al.*, 2001a) and genetic variation for these traits are very well documented (Kakani *et al.*, 2002, Prasad *et al.*, 2001a). The maximum percentage of pollen fertility, germination and pollen tube growth under high temperature are vital indicators of heat tolerant genotypes, therefore, suggested to deploy for screening in groundnut (Kakani *et al.* 2002).

## IDENTIFICATION OF QUANTITATIVE TRAIT LOCI (QTL)/GENES LINKED TO HEAT-STRESS TOLERANCE

Recent developments in genomic tools have resulted in the utilization of genomics-assisted trait dissection in several crop plants including groundnut. Several studies have reported QTLs for resistance to various biotic stresses (Pandey *et al.*, 2017), and tolerance to abiotic stresses in groundnut (Gautami *et al.*, 2012). The QTLs for heat-stress tolerance are reported in different crop species such as wheat (Paliwal *et al.*, 2012), rice (Ye *et al.*, 2012), maize (Frova and Sari-Gorla, 1994) *etc.*

In many crops, heat-stress tolerance associated traits are found to be governed by QTLs. For example, plasma membrane stability has been successfully used as an indirect measure of selection for heat-stress tolerance in wheat (Blum *et al.*, 2001), sorghum (Marcum, 1998) and barley (Wahid and Shabbir, 2005). In rice, QTLs associated with spikelet fertility under high temperature was mapped with 12.6-17.6% phenotypic variance explained (PVE) (Ye *et al.*, 2012). Pollen viability under heat stress was studied in tomato with PVE of 36% (Xu *et al.*, 2017). The QTLs for yield under heat stress associated traits (filled pods, total number of seeds, grain yield per plot and percent pod setting) were mapped in chickpea with 50% PVE (Paul *et al.*, 2018). However, literature information on the identification of genes/QTLs governing heat-stress tolerance in groundnut is non-existent. With a recent focus on developing climate-smart groundnut cultivars, it is expected that more information will be available on QTLs for heat-stress tolerance surrogate traits in groundnut in the near future. At International Crops Research Institute for the Semi-arid Tropics (ICRISAT) based at Patancheru, India, one recombinant inbred line (RIL) population was developed by crossing JL 24 (heat-stress susceptible) and 55-437 (heat-stress tolerant) genotypes for QTL mapping. This RIL population was evaluated under heat stress at three different locations in India and the QTL mapping for heat-stress tolerance is under progress.

## OMICS TOOLS FOR HEAT TOLERANCE IN GROUNDNUT

The different platforms of molecular tools such as sequencing, genome-wide association mapping, microarray, genome editing, transgenic technology, *etc.* are available to study the molecular mechanism of heat-stress tolerance in groundnut. But, no literature is available till date on molecular characterization of genes/QTLs that regulate this process in groundnut. The sequencing of cultivated groundnut and its diploid progenitors done in the recent past has provided critical insights into the genomes (Bertioli *et al.*, 2016; Zhuang *et al.*, 2019). With this development, it is expected that more molecular related information will be available for heat tolerance in groundnut.

## FUTURE OF BREEDING HEAT-STRESS TOLERANT GROUNDNUT

Groundnut is an important food crop with a significant contribution to protein and energy to the global population. With about 25% protein in the kernels, groundnut is valuable to meet the future emerging global demand for plant based proteins. On the other hand, increasing the yield of food crops including groundnut is important to meet the food needs of the fast increasing global human population which is expected to reach 10 billion in the next 30 years[1]. Tremendous yield gains were achieved so far in most of the crops to meet the food needs of the growing population. However, the current rate of yield gains for major cereals (wheat, rice and maize) and legumes (soybean, cowpea and groundnut) are insufficient to meet future demand to feed the 10 billion people by 2050 (Ray *et al.*, 2012, 2013). The average global annual growth rates of area, production and yield of groundnut from 1961 to 2016 were 0.7%, 2.2% and 1.5%, respectively (Nigam *et al.*, Unpublished). Africa with ~14.00 m ha area and ~13.50 m t of production, and Asia with 11.50 m ha area and 27.00 m t of production, together, account for 95 and 90% global groundnut area and production, respectively. The yield gains in many Asian and African countries are poor compared to other regions of the world due to several biotic and abiotic stresses. Heat-stress is one of the major abiotic stress which reduces the yield across SAT regions of the globe through limiting the crop growth, flowering, pod set and development along with the reduced nutritional quality of the product. The impact projections of 1.5 °C rise in global warming on human and natural ecosystems especially on crops showed a drastic reduction in yields of different crops across the SAT regions of the world particularly West Africa (Hoegh-Guldberg *et al.*, 2018). Several studies identified mechanisms to reduce the effects of heat stress during flowering in rice that includes heat avoidance through transpiration cooling (Julia and Dingkuhn, 2013), heat escape through early morning flowering (Ishimaru *et al.*, 2010; Julia and Dingkuhn, 2012; Hirabayashi *et al.*, 2014), and heat-tolerance through resilient reproductive processes (Jagadish *et al.*, 2009). These mechanisms and/or traits are less explored in other cereals and legumes. Hence, the mechanism of heat-tolerance in groundnut needs to be understood in order to breed resilient groundnut cultivars with higher yield under the changing climate scenario. The following strategies can be adopted to breed for heat-stress tolerant groundnut genotypes.

1. Screening of genotypes or breeding populations at select locations that are more prone to high temperatures can be used as selection criteria. As the frequency, duration and timing of heat stress can vary from location to location this strategy will be useful to identify stable as well as high performing genotypes.

2. Altering the sowing date or use of short-duration varieties that mature in 90-100 days can help escape periods of high temperature.

3. Development of multi-parent populations such as nested mapping association, multi-parent advanced generation intercross can be used to map alleles/QTLs linked to heat-tolerance, as well as identify lines that exhibit tolerance to high temperature stress. These can be later used as parent in breeding programs or released as a variety based on performance in different trials.

4. Earlier studies on the response of groundnut genotypes to heat stress both under field and controlled conditions have indicated that genotypic variations exist with respect to tolerance and these need to be utilized in breeding programs.

5. Wild relatives have more adaptation to different kinds of abiotic stresses. The identification and utilization of useful heat-stress tolerance surrogate traits/genes from wealthy wild relatives into breeding programs could sustain genetic gains through increased resilience to heat stress (Pradhan *et al.*, 2012; Pradhan and Prasad, 2015).

6. Emerging techniques such as speed breeding to reduce generation time and improve upon the genetic gains needs to be included in breeding strategies.

## CONCLUSION

Groundnut is an important oilseed, food, fooder and feed crop mainly cultivated in the SAT regions of Asia and Africa. The temperature regimes in the SATs are already close to or above the optimum levels (Singh *et al.*, 2014). Heat stress which results from short or long-term exposure to high temperatures above the threshold, causes negative effects on the overall growth and development of plants with a major effect on the reproductive phase. Even a mild stress at anthesis or grain filling phase can substantially reduce the crop yield. However, the period of heat temperatures and its diurnal timing will be the determining factor to indicate the extent of heat stress and its impacts on crop plants under specific climatic zones (Hasanuzzaman *et al.*, 2013a). Studies on groundnut indicates that heat stress mainly affects photosynthetic machinery, membrane stability, flowering, pollen production and viability, root nodulation, pod set and seed filling through accumulation of different metabolites along with alterations in metabolic pathways and processes.

The ability to withstand heat-stress greatly varies from species to species. The genetic approaches are being successfully exploited and deployed to minimize the negative effects of heat-stress in many crops. However, a detailed study on the physiological, biochemical and molecular mechanisms involved in the response and adaptation of groundnut to heat-stress, and the mechanism underlying the development of heat-stress tolerance needs better understanding. There is an urgent need to identify and characterize alleles/genes/QTLs for heat-tolerance in groundnut, and develop markers for utilization in the breeding programs. The availability of the genome of the sequence of cultivated groundnut and recent efforts in development and precise phenotyping of biparental and GS training populations would result in a better understanding of this complex trait and development of robust genomic tools for breeding programs in near future. Furthermore, it is important to keep an eye on upcoming climate change predictions which vary geographically. This information can guide breeding programs in updating the protocols for screening, traits for selection and identification of genotypes adapted to future climatic conditions. Apart from genetic approaches, adopting proper agronomic management practices such as altering the timing of sowing, irrigation management and selection of stress tolerance cultivar can to some extent, help minimize the adverse effects of heat-stress and improving crop productivity and food security in SAT regions of Asia and Africa.

## CONSENT FOR PUBLICATION

Not applicable.

## CONFLICT OF INTEREST

The author declares that there is no conflict of interest in this chapter.

## ACKNOWLEDGEMENTS

The authors acknowledge CGIAR's Research Program on Grain Legumes and Dryland Cereals (CRP-GLDC) (http://gldc.cgiar.org/) under which the groundnut improvement work is conducted at ICRISAT.

## NOTES

[1] 1https://www.un.org/development/desa/en/news/population/world-populatio--prospects-2017.html

## REFERENCES

Akbar, A., Singh Manohar, S., Tottekkaad Variath, M., Kurapati, S., Pasupuleti, J. (2017). Efficient partitioning of assimilates in stress-tolerant groundnut genotypes under high-temperature stress. *Agronomy (Basel).,  7*(2), 30.
[http://dx.doi.org/10.3390/agronomy7020030]

Awal, M.A., Ikeda, T. (2002). Effects of changes in soil temperature on seedling emergence and phenological development in field-grown stands of peanut (*Arachis hypogaea* L.). *Environ. Exp. Bot.,  47*(2), 101-113.
[http://dx.doi.org/10.1016/S0098-8472(01)00113-7]

Awal, M.A., Ikeda, T., Itoh, R. (2003). The effect of soil temperature on source-sink economy in peanut (*Arachis hypogaea*). *Environ. Exp. Bot.,  50*(1), 41-50.
[http://dx.doi.org/10.1016/S0098-8472(02)00111-9]

Babitha, M., Sudhakar, P., Latha, P., Reddy, P.V., Vasanthi, R.P. (2006). Screening of groundnut genotypes for high water use efficiency and temperature tolerance. *Indian J. Plant. Physiol.,  11*(1), 63.

Bäurle, I. (2016). Plant heat adaptation: Priming in response to heat stress. *F1000 Res.,  5*, 5.
[http://dx.doi.org/10.12688/f1000research.7526.1] [PMID: 27134736]

Bertioli, D.J., Cannon, S.B., Froenicke, L., Huang, G., Farmer, A.D., Cannon, E.K., Liu, X., Gao, D., Clevenger, J., Dash, S., Ren, L., Moretzsohn, M.C., Shirasawa, K., Huang, W., Vidigal, B., Abernathy, B., Chu, Y., Niederhuth, C.E., Umale, P., Araújo, A.C., Kozik, A., Kim, K.D., Burow, M.D., Varshney, R.K., Wang, X., Zhang, X., Barkley, N., Guimarães, P.M., Isobe, S., Guo, B., Liao, B., Stalker, H.T., Schmitz, R.J., Scheffler, B.E., Leal-Bertioli, S.C., Xun, X., Jackson, S.A., Michelmore, R., Ozias-Akins, P. (2016). The genome sequences of *Arachis duranensis* and *Arachis ipaensis*, the diploid ancestors of cultivated peanut. *Nat. Genet.,  48*(4), 438-446.
[http://dx.doi.org/10.1038/ng.3517] [PMID: 26901068]

Bita, C.E., Gerats, T. (2013). Plant tolerance to high temperature in a changing environment: Scientific fundamentals and production of heat stress-tolerant crops. *Front. Plant Sci.,  4*, 273.
[http://dx.doi.org/10.3389/fpls.2013.00273] [PMID: 23914193]

Blum, A., Klueva, N. (2001). Nguyen heat. Wheat cellular thermotolerance is related to yield under heat stress. *Euphytica.,  117*(2), 117-123.
[http://dx.doi.org/10.1023/A:1004083305905]

Boote, K.J., Allen, L.H., Prasad, P.V., Baker, J.T., Gesch, R.W., Snyder, A.M., Pan, D., Thomas, J.M. (2005). Elevated temperature and $CO_2$ impacts on pollination, reproductive growth, and yield of several globally important crops. *Nogyo Kisho.,  60*(5), 469-474.
[http://dx.doi.org/10.2480/agrmet.469]

Chakraborty, K., Bishi, S.K., Singh, A.L., Zala, P.V., Mahatma, M.K., Kalariya, K.A., Jat, R.A. (2018). Rapid induction of small heat shock proteins improves physiological adaptation to high temperature stress in peanut. *J. Agron. Crop Sci.,  204*(3), 285-297.

[http://dx.doi.org/10.1111/jac.12260]

Chakraborty, K., Singh, A.L., Kalariya, K.A., Goswami, N. (2015). Physiological responses of peanut (*Arachis hypogaea* L.) cultivars to water deficit stress: status of oxidative stress and antioxidant enzyme activities. *Acta Bot. Croat., 74*(1), 123-142.
[http://dx.doi.org/10.1515/botcro-2015-0011]

Chauhan, Y.S., Senboku, T. (1997). Evaluation of groundnut genotypes for heat tolerance 1. *Ann. Appl. Biol., 131*(3), 481-489.
[http://dx.doi.org/10.1111/j.1744-7348.1997.tb05175.x]

Cox, F.R. (1979). Effect of temperature treatment on peanut vegetative and fruit growth. *Peanut Sci., 6*(1), 14-17.
[http://dx.doi.org/10.3146/i0095-3679-6-1-4]

Craufurd, P.Q., Prasad, P.V., Kakani, V.G., Wheeler, T.R., Nigam, S.N. (2003). Heat tolerance in groundnut. *Field Crops Res., 80*(1), 63-77.
[http://dx.doi.org/10.1016/S0378-4290(02)00155-7]

Craufurd, P.Q., Prasad, P.V., Summerfield, R.J. (2002). Dry matter production and rate of change of harvest index at high temperature in peanut. *Crop Sci., 42*(1), 146-151.
[http://dx.doi.org/10.2135/cropsci2002.1460] [PMID: 11756265]

Devasirvatham, V., Tan, D.K., Trethowan, R.M. (2016). Breeding strategies for enhanced plant tolerance to heat stress. *Advances in Plant Breeding Strategies: Agronomic, Abiotic and Biotic Stress Traits.* (pp. 447-469). Cham: Springer.
[http://dx.doi.org/10.1007/978-3-319-22518-0_12]

Dixon, J.A., Gibbon, D.P., Gulliver, A. (2001). Farming systems and poverty: Improving farmers' livelihoods in a changing world. *Food. & Agricul. Org.*

Dreccer, M.F., Fainges, J., Sadras, V. Association between wheat yield and temperature in south-eastern Australia. *17th Australian Agronomy Conference,.* (2015). Hobart, Australia: Australian Society of Agronomy Inc. 559-561.

Dreyer, J., Duncan, W.G., McCloud, D.E. (1981). Fruit temperature, growth rates, and yield of peanuts 1. *Crop Sci., 21*(5), 686-688.
[http://dx.doi.org/10.2135/cropsci1981.0011183X002100050013x]

Driedonks, N., Rieu, I., Vriezen, W.H. (2016). Breeding for plant heat tolerance at vegetative and reproductive stages. *Plant Reprod., 29*(1-2), 67-79.
[http://dx.doi.org/10.1007/s00497-016-0275-9] [PMID: 26874710]

Fischer, G., Shah, M., Tubiello, F.N., van Velhuizen, H. (2005). Socio-economic and climate change impacts on agriculture: An integrated assessment, 1990-2080. *Philos. Trans. R. Soc. Lond. B Biol. Sci., 360*(1463), 2067-2083.
[http://dx.doi.org/10.1098/rstb.2005.1744] [PMID: 16433094]

Fischer, R.A., Maurer, R. (1978). Drought resistance in spring wheat cultivars. I. Grain yield responses. *Aust. J. Agric. Res., 29*(5), 897-912.
[http://dx.doi.org/10.1071/AR9780897]

Frimpong, A. (2004). Characterization of groundnut (*Arachis hypogaea* L.) in Northern Ghana. *Pak. J. Biol. Sci., 7*(5), 838-842.
[http://dx.doi.org/10.3923/pjbs.2004.838.842]

Frova, C., Sari-Gorla, M. (1994). Quantitative trait loci (QTLs) for pollen thermotolerance detected in maize. *Mol. Gen. Genet., 245*(4), 424-430.
[http://dx.doi.org/10.1007/BF00302254] [PMID: 7808391]

Fu, G., Feng, B., Zhang, C., Yang, Y., Yang, X., Chen, T., Zhao, X., Zhang, X., Jin, Q., Tao, L. (2016). Heat stress is more damaging to superior spikelets than inferiors of rice (*Oryza sativa* L.) due to their different organ temperatures. *Front. Plant Sci., 7*, 1637.

[http://dx.doi.org/10.3389/fpls.2016.01637] [PMID: 27877180]

Gangappa, E., Ravi, K., Kumar, G.N.V. (2006). Evaluation of groundnut (*Arachis hypogaea* L.) genotypes for temperature tolerance based on temperature induction response (TIR) technique. *Indian J. Genetics., 66*(2), 127-130.

Gautami, B., Pandey, M.K., Vadez, V., Nigam, S.N., Ratnakumar, P., Krishnamurthy, L., Radhakrishnan, T., Gowda, M.V., Narasu, M.L., Hoisington, D.A., Knapp, S.J., Varshney, R.K. (2012). Quantitative trait locus analysis and construction of consensus genetic map for drought tolerance traits based on three recombinant inbred line populations in cultivated groundnut (*Arachis hypogaea* L.). *Mol. Breed., 30*(2), 757-772. [http://dx.doi.org/10.1007/s11032-011-9660-0] [PMID: 22924017]

Golombek, S.D., Johansen, C. (1997). Effect of soil temperature on vegetative and reproductive growth and development in three Spanish genotypes of peanut (*Arachis hypogaea* L.). *Peanut Sci., 24*(2), 67-72. [http://dx.doi.org/10.3146/i0095-3679-24-2-1]

Hall, A.E. (2004). Breeding for adaptation to drought and heat in cowpea. *Eur. J. Agron., 21*(4), 447-454. [http://dx.doi.org/10.1016/j.eja.2004.07.005]

Hamidou, F., Halilou, O., Vadez, V. (2013). Assessment of groundnut under combined heat and drought stress. *J. Agron. Crop Sci., 199*(1), 1-1. [http://dx.doi.org/10.1111/j.1439-037X.2012.00518.x]

Hare, P.D., Cress, W.A., Van Staden, J. (1998). Dissecting the roles of osmolyte accumulation during stress. *Plant Cell Environ., 21*(6), 535-553. [http://dx.doi.org/10.1046/j.1365-3040.1998.00309.x]

Hasanuzzaman, M., Gill, S.S., Fujita, M. (2013). Physiological role of nitric oxide in plants grown under adverse environmental conditions. *Plant Acclimation to Environmental Stress.*, New York, NY: Springer. pp. 269-322. [http://dx.doi.org/10.1007/978-1-4614-5001-6_11]

Hasanuzzaman, M., Hossain, M.A., da Silva, J.A., Fujita, M. (2012). Plant response and tolerance to abiotic oxidative stress: Antioxidant defense is a key factor. *Crop Stress and its Management: Perspectives and Strategies.* Dordrecht: Springer. pp. 261-315.

Hasanuzzaman, M., Hossain, M.A., Fujita, M. (2010). Selenium in higher plants: Physiological role, antioxidant metabolism and abiotic stress tolerance. *J. Plant Sci., 5*(4), 354-375. [http://dx.doi.org/10.3923/jps.2010.354.375]

Hasanuzzaman, M., Nahar, K., Alam, M.M., Roychowdhury, R., Fujita, M. (2013). Physiological, biochemical, and molecular mechanisms of heat stress tolerance in plants. *Int. J. Mol. Sci., 14*(5), 9643-9684. [http://dx.doi.org/10.3390/ijms14059643] [PMID: 23644891]

Hirabayashi, H., Sasaki, K., Kambe, T., Gannaban, R.B., Miras, M.A., Mendioro, M.S., Simon, E.V., Lumanglas, P.D., Fujita, D., Takemoto-Kuno, Y., Takeuchi, Y., Kaji, R., Kondo, M., Kobayashi, N., Ogawa, T., Ando, I., Jagadish, K.S., Ishimaru, T. (2015). qEMF3, a novel QTL for the early-morning flowering trait from wild rice, *Oryza officinalis*, to mitigate heat stress damage at flowering in rice, *O. sativa. J. Exp. Bot., 66*(5), 1227-1236. [http://dx.doi.org/10.1093/jxb/eru474] [PMID: 25534925]

Hoegh-Guldberg, O., Jacob, D., Taylor, M., Bindi, M., Brown, S., Camilloni, I., Diedhiou, A., Djalante, R., Ebi, K., Engelbrecht, F., Guiot, J. (2018). Impacts of 1.5 °C global warming on natural and human systems. In: Global warming of 1.5° C. *An IPCC Special Report.*, 175-311. IPCC Secretariat.

Houghton, JT, Ding, YD, Griggs, DJ, Noguer, M, van der Linden, PJ, Dai, X, Maskell, K, Johnson, CA (2001). Climate change 2001: *The Scientific Basis.* The Press Syndicate of the University of Cambridge.

Howden, S.M., Soussana, J.F., Tubiello, F.N., Chhetri, N., Dunlop, M., Meinke, H. (2007). Adapting agriculture to climate change. *Proc. Natl. Acad. Sci. USA, 104*(50), 19691-19696. [http://dx.doi.org/10.1073/pnas.0701890104] [PMID: 18077402]

Ishimaru, T., Hirabayashi, H., Ida, M., Takai, T., San-Oh, Y.A., Yoshinaga, S., Ando, I., Ogawa, T., Kondo,

M. (2010). A genetic resource for early-morning flowering trait of wild rice *Oryza officinalis* to mitigate high temperature-induced spikelet sterility at anthesis. *Ann. Bot., 106*(3), 515-520.
[http://dx.doi.org/10.1093/aob/mcq124] [PMID: 20566680]

Jagadish, S.V., Muthurajan, R., Oane, R., Wheeler, T.R., Heuer, S., Bennett, J., Craufurd, P.Q. (2010). Physiological and proteomic approaches to address heat tolerance during anthesis in rice (*Oryza sativa* L.). *J. Exp. Bot., 61*(1), 143-156.
[http://dx.doi.org/10.1093/jxb/erp289] [PMID: 19858118]

Jain, M., Prasad, P.V., Boote, K.J., Hartwell, A.L., Jr, Chourey, P.S. (2007). Effects of season-long high temperature growth conditions on sugar-to-starch metabolism in developing microspores of grain sorghum (*Sorghum bicolor* L. Moench). *Planta., 227*(1), 67-79.
[http://dx.doi.org/10.1007/s00425-007-0595-y] [PMID: 17680267]

Janila, P., Nigam, S.N. (2013). Phenotyping for groundnut (*Arachis hypogaea* L.) improvement. *Phenotyping for Plant Breeding.* (pp. 129-167). New York, NY: Springer.

Julia, C., Dingkuhn, M. (2013). Predicting temperature induced sterility of rice spikelets requires simulation of crop-generated microclimate. *Eur. J. Agron., 49*, 50-60.
[http://dx.doi.org/10.1016/j.eja.2013.03.006]

Julia, C., Dingkuhn, M. (2012). Variation in time of day of anthesis in rice in different climatic environments. *Eur. J. Agron., 43*, 166-174.
[http://dx.doi.org/10.1016/j.eja.2012.06.007]

Kakani, V.G., Prasad, P.V., Craufurd, P.Q., Wheeler, T.R. (2002). Response of *in vitro* pollen germination and pollen tube growth of groundnut (*Arachis hypogaea* L.) genotypes to temperature. *Plant Cell Environ., 25*(12), 1651-1661.
[http://dx.doi.org/10.1046/j.1365-3040.2002.00943.x]

Kakani, V.G., Wheeler, T.R., Craufurd, P.Q., Rachaputi, R.C. (2015). Effect of high temperature and water stress on groundnuts under field conditions. *In Combined Stresses in Plants.,* Springe.Cham.x: 159-180.
[http://dx.doi.org/10.1007/978-3-319-07899-1_8]

Kaushal, N., Bhandari, K., Siddique, K.H., Nayyar, H. (2016). Food crops face rising temperatures: An overview of responses, adaptive mechanisms, and approaches to improve heat tolerance. *Cogent Food Agric., 2*(1), 1134380.
[http://dx.doi.org/10.1080/23311932.2015.1134380]

Ketring, DL (1985). Physiological response of groundnut to temperature and water deficits—breeding implications. *Agrometeorol. Groundnut.,* 21-135.

Ketring, D.L. (1979). Physiology of Oil Seeds. VIII. Germination of Peanut Seeds Exposed to Subfreezing Temperatures while Drying in the Windrow. *Peanut Sci., 6*(2), 80-83.
[http://dx.doi.org/10.3146/i0095-3679-6-2-4]

Ketring, D.L. (1984). Temperature Effects on Vegetative and Reproductive Development of Peanut1, 2. *Crop Sci., 24*(5), 877-882.
[http://dx.doi.org/10.2135/cropsci1984.0011183X002400050012x]

Kishinevsky, B.D., Sen, D., Weaver, R.W. (1992). Effect of high root temperature on Bradyrhizobium-peanut symbiosis. *Plant Soil., 143*(2), 275-282.
[http://dx.doi.org/10.1007/BF00007883]

Kishor, P.K., Sangam, S., Amrutha, R.N., Laxmi, P.S., Naidu, K.R., Rao, K.R., Rao, S., Reddy, K.J., Theriappan, P., Sreenivasulu, N. (2005). Regulation of proline biosynthesis, degradation, uptake and transport in higher plants: its implications in plant growth and abiotic stress tolerance. *Curr. Sci., 88*(3), 424-438.

Krasensky, J., Jonak, C. (2012). Drought, salt, and temperature stress-induced metabolic rearrangements and regulatory networks. *J. Exp. Bot., 63*(4), 1593-1608.
[http://dx.doi.org/10.1093/jxb/err460] [PMID: 22291134]

Kumar, R., Singh, A.K., Lavania, D., Siddiqui, M.H., Al-Whaibi, M.H., Grover, A. (2016). Expression

analysis of ClpB/Hsp100 gene in faba bean (*Vicia faba* L.) plants in response to heat stress. *Saudi J. Biol. Sci.,  23*(2), 243-247.
[http://dx.doi.org/10.1016/j.sjbs.2015.03.006] [PMID: 26981006]

Kumar, U, Singh, P, Boote, KJ (2012). Effect of climate change factors on processes of crop growth and development and yield of groundnut (*Arachis hypogaea* L.). *Advances in Agronomy.,  116*, 41-69. Academic Press.

Leong, S.K., Ong, C.K. (1983). The influence of temperature and soil water deficit on the development and morphology of groundnut (*Arachis hypogaea* L.). *J. Exp. Bot.,  34*(11), 1551-1561.
[http://dx.doi.org/10.1093/jxb/34.11.1551]

Lie, T.A. (1974). Environmental effects on nodulation and symbiotic nitrogen fixation. The biology of Nitrogen fixation. *Front. Biol.,  33*, 555-582.

Liu, X., Huang, B. (2000). Heat stress injury in relation to membrane lipid peroxidation in creeping bentgrass. *Crop Sci.,  40*(2), 503-510.
[http://dx.doi.org/10.2135/cropsci2000.402503x]

Marcum, K.B. (1998). Cell membrane thermostability and whole-plant heat tolerance of Kentucky bluegrass. *Crop Sci.,  38*(5), 1214-1218.
[http://dx.doi.org/10.2135/cropsci1998.0011183X003800050017x]

Martineau, J.R., Specht, J.E., Williams, J.H., Sullivan, C.Y. (1979). Temperature tolerance in soybeans. I. Evaluation of a technique for assessing cellular membrane thermostability 1. *Crop Sci.,  19*(1), 75-78.
[http://dx.doi.org/10.2135/cropsci1979.0011183X001900010017x]

Mills, W.T. (1964). Heat unit system for predicting optimum peanut-harvesting time. *Trans. ASAE.,  7*(3), 307-0309.
[http://dx.doi.org/10.13031/2013.40765]

Mohamed, H.A., Clark, J.A., Ong, C.K. (1988). Genotypic differences in the temperature responses of tropical crops: I. Germination characteristics of groundnut (*Arachis hypogaea* L.) and pearl millet (*Pennisetum typhoides* S. & H.). *J. Exp. Bot.,  39*(8), 1121-1128. a
[http://dx.doi.org/10.1093/jxb/39.8.1121]

Mohamed, H.A., Clark, J.A., Ong, C.K. (1988). Genotypic Differences in the Temperature Responses of Tropical Crops: II. Seedling emergence and leaf growth of groundnut (*Arachis hypogaea* L.) and pearl millet (*Pennisetum typhoides* S. & H.). *J. Exp. Bot.,  39*(8), 1129-1135. b
[http://dx.doi.org/10.1093/jxb/39.8.1129]

Nagarajan, S., Jagadish, S.V., Prasad, A.H., Thomar, A.K., Anand, A., Pal, M., Agarwal, P.K. (2010). Local climate affects growth, yield and grain quality of aromatic and non-aromatic rice in northwestern India. *Agric. Ecosyst. Environ.,  138*(3-4), 274-281.
[http://dx.doi.org/10.1016/j.agee.2010.05.012]

Naidu, G., Motagi, B., Gowda, M.C.B. (2017). Genetic variability for induced thermotolerance in groundnut (*Arachis hypogaea* L.) germplasm. *Electron. J. Plant Breed.,  8*(4), 1191-1196.
[http://dx.doi.org/10.5958/0975-928X.2017.00172.7]

Nautiyal, P.C., Rajgopal, K., Zala, P.V., Pujari, D.S., Basu, M., Dhadhal, B.A., Nandre, B.M. (2008). Evaluation of wild *Arachis* species for abiotic stress tolerance: I. Thermal stress and leaf water relations. *Euphytica,  159*(1-2), 43-57.
[http://dx.doi.org/10.1007/s10681-007-9455-x]

Ni, Z., Li, H., Zhao, Y., Peng, H., Hu, Z., Xin, M., Sun, Q. (2018). Genetic improvement of heat tolerance in wheat: recent progress in understanding the underlying molecular mechanisms. *Crop J.,  6*(1), 32-41.
[http://dx.doi.org/10.1016/j.cj.2017.09.005]

Ntare, B.R., Williams, J.H., Dougbedji, F. (2001). Evaluation of groundnut genotypes for heat tolerance under field conditions in a Sahelian environment using a simple physiological model for yield. *J. Agric. Sci.,  136*(1), 81-88.

[http://dx.doi.org/10.1017/S0021859600008583]

Nyarko, G., Alderson, P.G., Craigon, J., Murchie, E., Sparkes, D.L. (2008). Comparison of cell membrane thermostability and chlorophyll fluorescence parameters for the determination of heat tolerance in ten cabbage lines. *J. Hortic. Sci. Biotechnol., 83*(5), 678-682.
[http://dx.doi.org/10.1080/14620316.2008.11512443]

Oakes, A.J. (1958). Pollen Behavior in the Peanut (*Arachis hypogaea* L.) 1. *Agron. J., 50*(7), 387-389.
[http://dx.doi.org/10.2134/agronj1958.00021962005000070011x]

Opio, P., Photchanachai, S. (2016). Heat stress influences dormancy in peanut seeds (*Arachis hypogea* L.). *South Western J. Hortic. Biol. Env., 7*(2), 127-137.

Ortiz, R., Sayre, K.D., Govaerts, B., Gupta, R., Subbarao, G.V., Ban, T., Hodson, D., Dixon, J.M., Ortiz-Monasterio, J.I., Reynolds, M. (2008). Climate change: can wheat beat the heat? *Agric. Ecosyst. Environ., 126*(1-2), 46-58.
[http://dx.doi.org/10.1016/j.agee.2008.01.019]

Padmavathi, T.A., Rao, D.M. (2013). Differential accumulation of osmolytes in 4 cultivars of peanut (*Arachis hypogaea* L.) under drought stress. *J. Crop Sci. Biotechnol., 16*(2), 151-159.
[http://dx.doi.org/10.1007/s12892-012-0102-2]

Paliwal, R., Röder, M.S., Kumar, U., Srivastava, J.P., Joshi, A.K. (2012). QTL mapping of terminal heat tolerance in hexaploid wheat (*T. aestivum* L.). *Theor. Appl. Genet., 125*(3), 561-575.
[http://dx.doi.org/10.1007/s00122-012-1853-3] [PMID: 22476874]

Pandey, M.K., Wang, H., Khera, P., Vishwakarma, M.K., Kale, S.M., Culbreath, A.K., Holbrook, C.C., Wang, X., Varshney, R.K., Guo, B. (2017). Genetic dissection of novel QTLs for resistance to leaf spots and tomato spotted wilt virus in peanut (*Arachis hypogaea* L.). *Front. Plant Sci., 8*, 25.
[http://dx.doi.org/10.3389/fpls.2017.00025] [PMID: 28197153]

Paul, P.J., Samineni, S., Thudi, M., Sajja, S.B., Rathore, A., Das, R.R., Khan, A.W., Chaturvedi, S.K., Lavanya, G.R., Varshney, R.K., Gaur, P.M. (2018). Molecular mapping of QTLs for heat tolerance in chickpea. *Int. J. Mol. Sci., 19*(8), 2166.
[http://dx.doi.org/10.3390/ijms19082166] [PMID: 30044369]

Paulsen, GM (1994). High temperature responses of crop plants. *Physiology and Determination of Crop Yield., Oct;1*, 365-89.
[http://dx.doi.org/10.2134/1994.physiologyanddetermination.c25]

Pradhan, G.P., Prasad, P.V., Fritz, A.K., Kirkham, M.B., Gill, B.S. (2012). High temperature tolerance in *Aegilops* species and its potential transfer to wheat. *Crop Sci., 52*(1), 292-304.
[http://dx.doi.org/10.2135/cropsci2011.04.0186]

Pradhan, G.P., Prasad, P.V. (2015). Evaluation of wheat chromosome translocation lines for high temperature stress tolerance at grain filling stage. *PLoS One., 10*(2), e0116620.
[http://dx.doi.org/10.1371/journal.pone.0116620] [PMID: 25719199]

Prasad, P.V., Boote, K.J., Thomas, J.M., Allen, L.H., Jr, Gorbet, D.W. (2006). Influence of soil temperature on seedling emergence and early growth of peanut cultivars in field conditions. *J. Agron. Crop Sci., 192*(3), 168-177.
[http://dx.doi.org/10.1111/j.1439-037X.2006.00198.x]

Prasad, P.V., Craufurd, P.Q., Kakani, V.G., Wheeler, T.R., Boote, K.J. (2001). Influence of high temperature during pre-and post-anthesis stages of floral development on fruit-set and pollen germination in peanut. *Funct. Plant Biol., 28*(3), 233-240. a
[http://dx.doi.org/10.1071/PP00127]

Prasad, P.V., Craufurd, P.Q., Summerfield, R.J., Wheeler, T.R. (2000). Effects of short episodes of heat stress on flower production and fruit-set of groundnut (*Arachis hypogaea* L.). *J. Exp. Bot., 51*(345), 777-784. a
[PMID: 10938870]

Prasad, P.V., Craufurd, P.Q., Summerfield, R.J. (2000). Effect of high air and soil temperature on dry matter

production, pod yield and yield components of groundnut. *Plant Soil., 222*(1-2), 231-239. b
[http://dx.doi.org/10.1023/A:1004793220787]

Prasad, P.V., Craufurd, P.Q., Summerfield, R.J. (1999). Fruit number in relation to pollen production and viability in groundnut exposed to short episodes of heat stress. *Ann. Bot. (Lond.), 84*(3), 381-386. a
[http://dx.doi.org/10.1006/anbo.1999.0926]

Prasad, P.V., Craufurd, P.Q., Summerfield, R.J. (2001). Response of groundnuts dependent on symbiotic and inorganic nitrogen to high air and soil temperatures. *J. Plant Nutr., 24*(4-5), 623-637. b
[http://dx.doi.org/10.1081/PLN-100103657]

Prasad, P.V., Craufurd, P.Q., Summerfield, R.J. (1999). Sensitivity of peanut to timing of heat stress during reproductive development. *Crop Sci., 39*(5), 1352-1357. b
[http://dx.doi.org/10.2135/cropsci1999.3951352x]

Prasad, P.V., Boote, K.J., Hartwell Allen Jr, L., Thomas, J.M. (2003). Superoptimal temperatures are detrimental to peanut (*Arachis hypogaea* L.) reproductive processes and yield at both ambient and elevated carbon dioxide. *Glob. Change Biol., 9*(12), 1775-1787.
[http://dx.doi.org/10.1046/j.1365-2486.2003.00708.x]

Rani, K.R., Chamundeswari, K., Usha, R. (2018). Screening of thermotolerant groundnut genotypes using temperature induction response–a novel approach to assess genetic variability. *Int. J. Pharma Bio Sci., 8*, 360-364.

Rasheed, R., Wahid, A., Farooq, M., Hussain, I., Basra, S.M. (2011). Role of proline and glycinebetaine pretreatments in improving heat tolerance of sprouting sugarcane (*Saccharum* sp.) buds. *Plant Growth Regul., 65*(1), 35-45.
[http://dx.doi.org/10.1007/s10725-011-9572-3]

Ray, D.K., Mueller, N.D., West, P.C., Foley, J.A. (2013). Yield trends are insufficient to double global crop production by 2050. *PLoS One., 8*(6), e66428.
[http://dx.doi.org/10.1371/journal.pone.0066428] [PMID: 23840465]

Ray, D.K., Ramankutty, N., Mueller, N.D., West, P.C., Foley, J.A. (2012). Recent patterns of crop yield growth and stagnation. *Nat. Commun., 3*, 1293.
[http://dx.doi.org/10.1038/ncomms2296] [PMID: 23250423]

Rebetzke, G.J., Rattey, A.R., Farquhar, G.D., Richards, R.A., Condon, A.T.G. (2012). Genomic regions for canopy temperature and their genetic association with stomatal conductance and grain yield in wheat. *Funct. Plant Biol., 40*(1), 14-33.
[http://dx.doi.org/10.1071/FP12184] [PMID: 32481083]

Roberts, E.H. (1988). Temperature and seed germination.

Sadat, S., Saeid, K.A., Bihamta, M.R., Torabi, S., Salekdeh, S.G., Ayeneh, G.A. (2013). Marker assisted selection for heat tolerance in bread wheat. *World Appl. Sci. J., 21*(8), 1181-1189.

Saffan, S.E. (2008). Effect of heat stress on phytochemical composition of peanut seedlings. *Res. J. Agric. Biol. Sci., 4*(2), 167-174.

Sakamoto, A., Murata, N. (2002). The role of glycine betaine in the protection of plants from stress: clues from transgenic plants. *Plant Cell Environ., 25*(2), 163-171.
[http://dx.doi.org/10.1046/j.0016-8025.2001.00790.x] [PMID: 11841661]

Savchenko, G.E., Klyuchareva, E.A., Abramchik, L.M., Serdyuchenko, E.V. (2002). Effect of periodic heat shock on the inner membrane system of etioplasts. *Russ. J. Plant Physiol., 49*(3), 349-359.
[http://dx.doi.org/10.1023/A:1015592902659]

Scafaro, A.P., Haynes, P.A., Atwell, B.J. (2010). Physiological and molecular changes in *Oryza meridionalis* Ng., a heat-tolerant species of wild rice. *J. Exp. Bot., 61*(1), 191-202.
[http://dx.doi.org/10.1093/jxb/erp294] [PMID: 19819927]

Schuster, W.S., Monson, R.K. (1990). An examination of the advantages of C3C4 intermediate

photosynthesis in warm environments. *Plant Cell Environ.,* *13*(9), 903-912.
[http://dx.doi.org/10.1111/j.1365-3040.1990.tb01980.x]

Selvaraj, M.G., Burow, G., Burke, J.J., Belamkar, V., Puppala, N., Burow, M.D. (2011). Heat stress screening of peanut *(Arachis hypogaea* L.) seedlings for acquired thermotolerance. *Plant Growth Regul.,* *65*(1), 83-91.
[http://dx.doi.org/10.1007/s10725-011-9577-y]

Semenov, M.A., Halford, N.G. (2009). Identifying target traits and molecular mechanisms for wheat breeding under a changing climate. *J. Exp. Bot.,* *60*(10), 2791-2804.
[http://dx.doi.org/10.1093/jxb/erp164] [PMID: 19487387]

Singh, D., Balota, M., Collakova, E., Isleib, T.G., Welbaum, G.E., Tallury, S.P. (2016). Heat stress related physiological and metabolic traits in peanut seedlings. *Peanut Sci.,* *43*(1), 24-35.
[http://dx.doi.org/10.3146/0095-3679-43.1.24]

Singh, P., Nedumaran, S., Ntare, B.R., Boote, K.J., Singh, N.P., Srinivas, K., Bantilan, M.C. (2014). Potential benefits of drought and heat tolerance in groundnut for adaptation to climate change in India and West Africa. *Mitig. Adapt. Strategies Glob. Change,* *19*(5), 509-529.
[http://dx.doi.org/10.1007/s11027-012-9446-7]

Sita, K., Sehgal, A., HanumanthaRao, B., Nair, R.M., Vara Prasad, P.V., Kumar, S., Gaur, P.M., Farooq, M., Siddique, K.H.M., Varshney, R.K., Nayyar, H. (2017). Food legumes and rising temperatures: Effects, adaptive functional mechanisms specific to reproductive growth stage and strategies to improve heat tolerance. *Front. Plant Sci.,* *8*, 1658.
[http://dx.doi.org/10.3389/fpls.2017.01658] [PMID: 29123532]

Soliman, W.S., Fujimori, M., Tase, K., Sugiyama, S.I. (2011). Oxidative stress and physiological damage under prolonged heat stress in C3 grass *Lolium perenne. Grassl. Sci.,* *57*(2), 101-106.
[http://dx.doi.org/10.1111/j.1744-697X.2011.00214.x]

Solomon, S, Manning, M, Marquis, M, Qin, D. (2007). Climate change 2007-the physical science basis: Working group I contribution to the fourth assessment report of the IPCC. *Cambridge University Press.* Sep 10.

Srinivasan, A., Takeda, H., Senboku, T. (1996). Heat tolerance in food legumes as evaluated by cell membrane thermostability and chlorophyll fluorescence techniques. *Euphytica,* *88*(1), 35-45.
[http://dx.doi.org/10.1007/BF00029263]

Taiz, L., Zeiger, E. (2006). *Plant Physiology..* Sunderland, MA: Sinauer Associates, Inc., Publishers.

Talwar, H.S., Takeda, H., Yashima, S., Senboku, T. (1999). Growth and photosynthetic responses of groundnut genotypes to high temperature. *Crop Sci.,* *39*(2), 460-466.
[http://dx.doi.org/10.2135/cropsci1999.0011183X0039000200027x]

Talwar, H.S., Yanagihara, S., Yajima, M., Hayashi, T. (1999).

Tripp, J., Mishra, S.K., Scharf, K.D. (2009). Functional dissection of the cytosolic chaperone network in tomato mesophyll protoplasts. *Plant Cell Environ.,* *32*(2), 123-133.
[http://dx.doi.org/10.1111/j.1365-3040.2008.01902.x] [PMID: 19154229]

Tsukaguchi, T., Kawamitsu, Y., Takeda, H., Suzuki, K., Egawa, Y. (2003). Water status of flower buds and leaves as affected by high temperature in heat-tolerant and heat-sensitive cultivars of snap bean (*Phaseolus vulgaris* L.). *Plant Prod. Sci.,* *6*(1), 24-27.
[http://dx.doi.org/10.1626/pps.6.24]

Usman, M.G., Rafii, M.Y., Ismail, M.R., Malek, M.A., Latif, M.A., Oladosu, Y. (2014). Heat shock proteins: functions and response against heat stress in plants. *Int. J. Sci. Tech. Res.,* *3*(11), 204-218.

Venkateswarlu, B., Ramesh, K. (1993). Cell membrane stability and biochemical response of cultured cells of groundnut under polyethylene glycol-induced water stress. *Plant Sci.,* *90*(2), 179-185.
[http://dx.doi.org/10.1016/0168-9452(93)90238-U]

Wahid, A., Close, T.J. (2007). Expression of dehydrins under heat stress and their relationship with water

Wahid, A., Close, T.J. (2007). Expression of dehydrins under heat stress and their relationship with water relations of sugarcane leaves. *Biol. Plant., 51*(1), 104-109.
[http://dx.doi.org/10.1007/s10535-007-0021-0]

Wahid, A., Shabbir, A. (2005). Induction of heat stress tolerance in barley seedlings by pre-sowing seed treatment with glycinebetaine. *Plant Growth Regul., 46*(2), 133-141.
[http://dx.doi.org/10.1007/s10725-005-8379-5]

Wahid, A. (2007). Physiological implications of metabolite biosynthesis for net assimilation and heat-stress tolerance of sugarcane (*Saccharum officinarum*) sprouts. *J. Plant Res., 120*(2), 219-228.
[http://dx.doi.org/10.1007/s10265-006-0040-5] [PMID: 17024517]

Wang, L.J., Fan, L., Loescher, W., Duan, W., Liu, G.J., Cheng, J.S., Luo, H.B., Li, S.H. (2010). Salicylic acid alleviates decreases in photosynthesis under heat stress and accelerates recovery in grapevine leaves. *BMC Plant Biol., 10*(1), 34.
[http://dx.doi.org/10.1186/1471-2229-10-34] [PMID: 20178597]

Wang, W., Vinocur, B., Shoseyov, O., Altman, A. (2004). Role of plant heat-shock proteins and molecular chaperones in the abiotic stress response. *Trends Plant Sci., 9*(5), 244-252.
[http://dx.doi.org/10.1016/j.tplants.2004.03.006] [PMID: 15130550]

Waraich, E.A., Ahmad, R., Halim, A., Aziz, T. (2012). Alleviation of temperature stress by nutrient management in crop plants: a review. *J. Soil Sci. Plant Nutr., 12*(2), 221-244.
[http://dx.doi.org/10.4067/S0718-95162012000200003]

Wheeler, T.R., Craufurd, P.Q., Ellis, R.H., Porter, J.R., Prasad, P.V. (2000). Temperature variability and the yield of annual crops. *Agric. Ecosyst. Environ., 82*(1-3), 159-167.
[http://dx.doi.org/10.1016/S0167-8809(00)00224-3]

Williams, J.H., Saxena, N.P. (1991). The use of non destructive measurement and physiological models of yield determination to investigate factors determining differences in seed yield between genotypes of "desi" chickpeas (*Cicer arietum*). *Ann. Appl. Biol., 119*(1), 105-112.
[http://dx.doi.org/10.1111/j.1744-7348.1991.tb04848.x]

Wynne, J.C., Sullivan, G.A. (1978). Effect of environment and cultivar on peanut seedling emergence. *Peanut Sci., 5*(2), 109-111.
[http://dx.doi.org/10.3146/i0095-3679-5-2-13]

Xu, J., Driedonks, N., Rutten, M.J.M., Vriezen, W.H., de Boer, G.J., Rieu, I. (2017). Mapping quantitative trait loci for heat tolerance of reproductive traits in tomato (*Solanum lycopersicum*). *Mol. Breed., 37*(5), 58.
[http://dx.doi.org/10.1007/s11032-017-0664-2] [PMID: 28479863]

Xu, S., Li, J., Zhang, X., Wei, H., Cui, L. (2006). Effects of heat acclimation pretreatment on changes of membrane lipid peroxidation, antioxidant metabolites, and ultrastructure of chloroplasts in two cool-season turfgrass species under heat stress. *Environ. Exp. Bot., 56*(3), 274-285.
[http://dx.doi.org/10.1016/j.envexpbot.2005.03.002]

Yang, S., Wang, F., Guo, F., Meng, J.J., Li, X.G., Dong, S.T., Wan, S.B. (2013). Exogenous calcium alleviates photoinhibition of PSII by improving the xanthophyll cycle in peanut (*Arachis hypogaea*) leaves during heat stress under high irradiance. *PLoS One, 8*(8), e71214.
[http://dx.doi.org/10.1371/journal.pone.0071214] [PMID: 23940721]

Ye, C., Argayoso, M.A., Redoña, E.D., Sierra, S.N., Laza, M.A., Dilla, C.J., Mo, Y., Thomson, M.J., Chin, J., Delaviña, C.B., Diaz, G.Q. (2012). Mapping QTL for heat tolerance at flowering stage in rice using SNP markers. *Plant Breed., 131*(1), 33-41.
[http://dx.doi.org/10.1111/j.1439-0523.2011.01924.x]

Young, J.C. (2010). Mechanisms of the Hsp70 chaperone system. *Biochem. Cell Biol., 88*(2), 291-300.
[http://dx.doi.org/10.1139/O09-175] [PMID: 20453930]

# CHAPTER 6

# Mungbean And High-Temperature Stress: Responses And Strategies To Improve Heat Tolerance

**Manu Priya[1], Aditya Pratap[2], Debjoti Sengupta[2], Kadambot H.M Siddique[3], N.P. Singh[2], Uday Jha[2]** and **Harsh Nayyar[1,\*]**

[1] *Department of Botany, Panjab University, Chandigarh, India*

[2] *Indian Institute of Pulses Research, Kanpur (U.P.), India*

[3] *The UWA Institute of Agriculture M082, Perth WA6009, Australia*

**Abstract:** Considering the current scenario of global climate change, high-temperature stress is becoming a major threat limiting crop yield and productivity of crops including mungbean (*Vigna radiata* L. Wilczeck), globally. Significant yield reduction in mungbean due to high-temperature stress, especially during the reproductive stage, has been observed by various researchers. Therefore, identification of heat-tolerant mungbean lines by using different selection criteria, based on field trials evaluating various yield traits, is urgently needed. An overview of different morpho-physiological responses of mungbean under heat stress may help in formulating appropriate strategies for improving its yield potential. In addition, identification and incorporation of appropriate management strategies may enhance the productivity and sustainability of mungbean worldwide. The key findings of this chapter include the effects of heat stress on growth, reproduction and physiology of mungbean growing at different agro-climatic zones. Further, effective approaches for managing heat stress such as selection and screening of available germplasm under field trials, application of exogenous thermo-protectants and well-integrated genetic and agronomic management methods, are also discussed to improve mungbean performance under heat stress. However, the implications of the above-mentioned techniques for heat stress management require deep insight into heat tolerance mechanisms, molecular breeding, and gene characterization methods.

**Keywords:** Breeding, Mungbean, Heat tolerance, High temperature, Legumes.

## INTRODUCTION

Rising temperatures and associated climatic disturbances are considered a serious

---
\* **Corresponding author Harsh Nayyar:** Department of Botany, Panjab University, Chandigarh, India; E-mail:nayarbot@pu.ac.in

**Uday C. Jha, Harsh Nayyar and Sanjeev Gupta (Eds.)**

threat to future agricultural output and food supply (Hanumantha Rao *et al.*, 2016). High temperature reduces agricultural yield and productivity of different crop species including grain legumes, either directly or indirectly (Basu *et al.*, 2009; Farooq *et al.*, 2018; Sita *et al.*, 2017). Currently, increased adversities of high-temperature stress on crop productivity are receiving considerable attention worldwide (Teixeira *et al.*, 2013). Previous studies have speculated that global food productivity needs to be doubled by the end of 2050 to meet the requirements of rising population and dietary shifts(Ohama *et al.*, 2017). Enhancing agricultural yields to keep pace with these increasing demands have been suggested as a favored solution to achieve this goal (Fedoroff *et al.*, 2010; Godfray *et al.*, 2010). During the current era of global warming, hot spells and warm days are likely to enhance both in intensity as well as frequency in many temperate, sub-tropical and tropical areas of the world in the coming future (Team *et al.*, 2014), which is expected to diminish the crop production. Further, in view of the complete reliance of humans on agricultural crops for meeting food demands, a deep insight regarding the sensitivity of food crops towards heat stress at different developmental stages is of principal importance (Kumar *et al.*, 2013). According to current climate model predictions, generally, the arid and semi-arid regions of the world represent the badly affected areas due to rising temperature (Vadez *et al.*, 2012; Teixeira *et al.*, 2013). Consequently, food crops, especially summer- grown crops including food legumes such as Mungbean, especially being cultivated in tropical areas, encounter frequent spells of heatwaves, coupled with soil moisture stress (Vadez *et al.*, 2012; Sita *et al.*, 2017; Farooq *et al.*, 2018). Concerning these issues, efforts are required to examine the heat sensitivity of summer crops and to develop strategies to overcome the devastating impacts of elevating temperatures. In combination with drought or other stresses, high temperature leads to world-wide extensive loss to agriculture (Mittler, 2006). Carbohydrate metabolism is also impaired due to the incongruity between photosynthesis and respiration (Ruan *et al.*, 2010). Further, intensification in heat-stress also inhibits membrane functionality and essential physiological processes, ultimately leading to cell death (Hatfield and Prueger, 2015). Aberrant metabolism due to the whole sequence of events finally ends up in the generation of reactive oxygen species (ROS) and toxic metabolites in the injured cells, which causes oxidative damage, protein denaturation, and DNA mutation(Van Breusegem *et al.*, 2001). Seed germination may also be completely inhibited depending upon plant species, intensity, and duration of the stress (Rasheed *et al.*, 2016). Heat stress also alters the stability, compartmentalization, content, and homeostasis of many molecules, especially plant growth regulators (Maestri *et al.*, 2002) Some other consequences include premature shedding of leaves, flowers, and fruits which produce unproductive tillers due to loss of entire crop cycles (Guo *et al.*, 2016).

In legumes, high temperature drastically affects various plant growth parameters (Hamada, 2001;Kumar *et al.,* 2013) such as photosynthetic efficiency (PSI and PSII), stability of thylakoid membranes (Gounaris *et al.,* 1984), electron transport channels (Srinivasan *et al.,* 1996; Sharkey, 2005), respiration (Kurets and Popov, 1998) and nitrogen fixing ability (Zahran, 1999). The adversity of constantly rising temperatures in plants is reported to be higher during the reproductive phase in comparison to the vegetative phase (Hall, 1992). Further, male reproductive structures are more vulnerable to heat than the female reproductive structures, which is evident from various heat stress studies (Dickson and Boettger 1984; Monterroso and Wien,1990; Young *et al.,* 2004). Recently, some efforts are being made in developing stress-tolerant varieties of legumes either by traditional breeding strategies or by molecular-assisted breeding methods (Varshney *et al.,* 2014; Pratap *et al.,* 2017; Mannur *et al.,* 2019). The global climatic changes are a great challenge for agriculture, especially; a progressive increase in temperature is a major concern for the crops. In temperate areas too, short and occasional increase in temperature of several degrees above the mean values reported for a certain season occur more and more often (Sgobba *et al.,* 2015). Among grain legumes, considerable yield losses have been reported in mungbean (a summer-grown legume) due to heat stress during flowering, which also affects both root and shoot growth and results in the poor crop quality (Kaur *et al.,* 2015; Sharma *et al.,* 2016). Heat stress negatively affects the pollen maturation, pollen viability, pollen germination and pollen tube growth (Sita *et al.,* 2017;Basu *et al.,* 2019). Exposure of plants to heat stress during the seed filling stage enhances senescence, decreased seed set, seed weight, and yield as reported in mungbean (Kaur *et al.,* 2015). High temperature also induces chlorosis, senescence and abscission in leaves, inhibits proliferation of roots and shoots and inhibits the yield potential (Hossain *et al.,* 2012).

## GROWING CONDITIONS AND STATUS OF MUNGBEAN CULTIVATION

Mungbean (*Vigna radiata* L. Wilczek) is a valuable pulse crop in several Asian countries including India (Dahiya *et al.,* 2015; Thirumaran and Seralathan). India is one of the largest producers and consumer of mungbean contributing up to 54% of the world's production (Sehrawat *et al.,* 2013). The crop is generally grown in summer and autumn in an optimum temperature range between 27-30°C and mostly cultivated in arid and semiarid tropics at altitudes below 2000m (Singh *et al.,* 2017). The plant is an annual food legume belonging to family Fabaceae and has an indispensable role in nutrition all over the world (Pratap *et al.,* 2017). Mungbean is an economical source of plant protein ranging from 22-27% and is the main component of a balanced diet (Biswash *et al.,* 2014). Mungbean is an erect to a semi-erect slightly pubescent herb, which is highly branched and is about 60 to 70 cm tall (Lambrides and Godwin, 2007). For a sterling yield of mungbean, appropriate rainfall is essential from flowering to the late podding (Singh *et al.,* 2017).

The annual production of mungbean grain is >3 million tons from an area of about 6.0 million ha. However, due to increased global demand of vegetable protein, along with market price, mungbean is now commercially cultivated in large demand-led farms (Keatinge *et al.,* 2011). Besides, it is also sown in small specific niches such as rabi-rice fallow in Peninsular India during November and January. In Bangladesh, the crop is grown in two seasons; during 'rabi' season in November and during the 'kharif' season starting in June (Reddy, 2009). Earlier, due to high photo- and thermo-period sensitivity, different cultivars were used for each cropping season. However, in recent times a number of largely photo and thermo-period tolerant cultivars have been developed which can be cultivated across different seasons and locations.

In India, the total coverage area of mungbean was 42.50 lakh hectares with a total production of 24.10 lakh tons with an average productivity of 567 kg/ha (Project Coordinator, MULLaRP Report, 2019). The maximum area and production were recorded in Rajasthan (45.24% & 51.82%), followed by Karnataka (9.90 and 6.01%) and Maharashtra (9.29% & 5.98%). Madhya Pradesh ranked forth with 7.79% area) and Odisha ranked 5th with 5.25% area under mungbean cultivation (Project Coordinator, MULLaRP Report, 2018-19). Maximum yield was recorded in the state of Punjab (845 kg/ha) followed by Jharkhand (704 kg/ha) and Andhra Pradesh (696 kg/ha) in the period of five year *i.e.,* 2012-2017. The National average yield was recorded as 567 kg/ha during 2018-19. Further, the lowest yield was recorded in the state of Karnataka (345 kg/ha) followed by Odisha (360 kg/ha) and Maharashtra (365 kg/ha) (Project Coordinator, MULLaRP Report, 2019). Mungbean plays an important role in various cropping systems, and in agriculture, due to its nitrogen-fixing ability, large biomass production, low water requirement, and short life span (Biswash *et al.,* 2014). However, the high variability in climatic conditions, including rising temperatures, unpredictable rainfall and uneven soil moisture distribution are limiting mungbean productivity during its cropping season (Singh *et al.,* 2017).

## THE NATURE OF HEAT STRESS AND CROP RESPONSE

Heat stress generally occurs in combination with drought, high solar irradiance and intense wind velocity, and all these factors together can aggravate damage even in fully-watered plants (Hall, 1992). The severity and occurrence of heat stress depends upon temperature regimes, plant genotype, and soil profile and water status in field conditions that vary greatly in different growing regions all over the world. Based upon the time of occurrence, duration and interaction of several factors, heat stress can be divided into acute and chronic, involving various coping strategies, adaptive means and ultimately, breeding approaches (Blum,1989). Acute heat stress is shorter in duration and may occur at any phase of crop growth cycle that often lowers yield. In contrast, chronic heat stress may be prevalent at every growth stage and usually causes significant yield losses and even crop failure. However, acute heat stress is more damaging than chronic heat stress in the summer-sown mungbean regions of south-east Asia including the Indian subcontinent. Being a warm-season crop, Mungbean experiences exposures of above-optimal temperature during its normal cultivation season, especially in late-growing conditions (Sharma *et al.*, 2016). The impacts are larger in late-sown crop and consequently its yield gets constrained markedly because of inhibition of vegetative growth, reproductive failures shown as drop in flowers and pod number, less pod filling and reduced seed size (Kaur *et al.*, 2015). Further, heat stress during reproductive stage is more prevalent than vegetative stage that reduces crop yield drastically (Sharma *et al.*, 2016). Yield reduction is associated with reduced pollen viability and aberrant fertilization events and under high temperature environment (Farooq *et al.*, 2011; Kaushal *et al.*, 2016). Rising atmospheric temperatures are highly detrimental for growth and physiological functions of various food crops, more so in mungbean (Cao *et al.*, 2011). Nevertheless, mungbean is a tropical pulse crop and flourishes better in all arid and semi-arid regions of the world; heat susceptibility indeed is a critical environmental restriction that hampers crop productivity adversely (Lateef *et al.*, 2018). In mungbean, reproductive tissues and their functions are more prone to heat stress, and rise in temperature during flowering may even lead to loss of entire crop cycles (Kaur *et al.*, 2015). Even an increase in temperature by few degrees changes crop cycle, and accelerates flower and pods abortion, malformation of fruits and poor grain filling(Kaur *et al.*, 2015; Kaushal *et al.*, 2016). The effects of heat stress on mungbean are not fully reported and need to be probed (Kaur *et al.*, 2015). Nevertheless, due to genotypic variability, different genotypes have a range of tolerance or resistance means that help them to withstand adverse effects of heat stress (Chauhan and Williams, 2018). The indeterminate flowering pattern and short season nature of mungbean cultivars are highly beneficial in sustaining cropping systems through escape mechanism though not for required traits to enhance yield under heat stress for commercial scale production (Kang *et al.*, 2014). The exploitation of available genetic variability in mungbean to develop higher yielding and heat tolerant cultivars still remain an element of significant interest among researchers (Chauhan and Williams, 2018). The key targets for mungbean improvement under heat stress include a maturity period of around 60 to 70 days, a significantly higher yield and synchronous maturity for ease in harvesting (Kim *et al.,* 2015). Some other targets include a compact canopy, resistance to other stress factors, photoperiod insensitivity, high harvest index (HI), and increased determinacy(Chauhan and Williams, 2018). The study of Lawn,(1989) showed that different varieties of mungbean have differential response to maximum and minimum temperatures. Further, photosynthetic events also showed high sensitivity towards elevated temperatures due to loss of chlorophyll and reduction in carbon fixation and assimilation (Sinsawat *et al.,* 2004). The work of Naveed *et al.,* (2015) revealed that optimum sowing date is an important trait for improving mungbean yield potential in varied agro-ecological realms of the world. Under heat stress, the trend of higher to lower harvest indices in mungbean may be due to shifts in optimum levels of both soil and air temperatures as well as precipitation during the growth cycle of crops (Naveed *et al.,* 2015). However, this trend is basically followed in some planting dates that enhances assimilate production and transport to reproductive sink such as grain (Naveed *et al.,* 2015). Warmer temperatures *i.e.,* >44/34°C affected net photosynthetic rate significantly at all developmental stages of mungbean (Chikukura *et al.,* 2017). A shift in the optimum seasonal temperatures by 8-10°C also shortens all phenological stages, particularly vegetative growth, leading to earlier maturity (50-57 DAS) in comparison to the control that attain maturity in 70-77 DAS (Chikukura *et al.,* 2017).

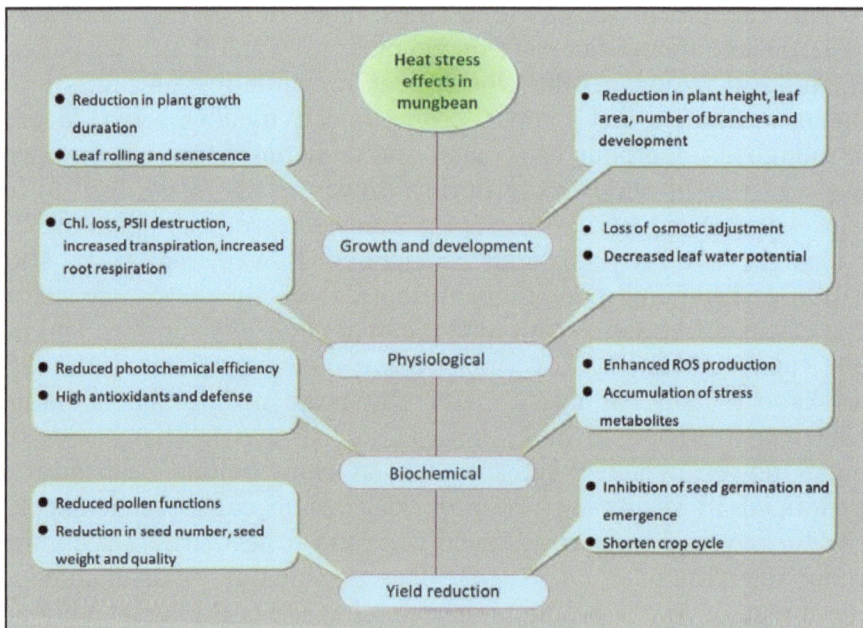

**Fig. (1).** Pictorial representation of morphological, physiological and biochemical responses of mungbean to combat harmful effects of heat stress.

# EFFECTS OF HEAT STRESS

## Growth And Phenology

A wide-array of studies documented high sensitivity of mungbean towards rising temperature(Jha *et al.*, 2017). Deleterious consequences of high temperature stress are prevalent on plant growth, development, as well as various physiological functions (Hanumantha Rao *et al.*, 2016). For instance, prolonged exposure of mungbean to high temperature may result in loss of vigor, consequently affecting seedling emergence and establishment (Fig. **1**) (Kumar *et al.*, 2011). In mungbean, high temperature *i.e.* 50°C for 10, 20, 30 minutes, reduced seed germination and vigor index significantly (Piramila *et al.*, 2012).

Fig. (2). Some symptoms on heat stress on mungbean. **a**: leaf scorching, **b**: aborted flowers, **c**: aborted pods, **d**: Shriveled pods.

The harmful effects of heat stress on vegetative growth include leaf senescence, chlorosis, necrosis, scorching, and abscission, reduced internode elongation and root and shoot growth inhibition (Fig. **1**) (Kaushal *et al.*, 2011). High temperature, especially >40/30°(max/min) causes growth inhibition and chlorosis in mungbean, which is associated with decrease in leaf water status and elevated oxidative stress, which was reported to be mitigated by exogenous application of ascorbic acid (Kumar *et al.*, 2011). Heat induced leaf scorching, leaf rolling and chlorosis has been also reported in mungbean plants by Sharma *et al.*, (2016). Moreover, acceleration in growth and phenology was observed, which further reduced leaf area, biomass production, number of flowers and pods in heat-stressed mungbean (Sharma *et al.*, 2016). Some other consequences of heat stress on Mungbean were leaf curling, leaf wilting, yellowing and blackening of leaves, reduction in plant height, and number of leaves, branches and biomass (Fig. **2a**) (Kaur *et al.*, 2015).

**Heat Stress And Reproductive Development**

Although plants have inherent ability to sustain their metabolism and vegetative growth even under varied temperature regimes, the reproductive growth showed considerable sensitivity towards warmer temperatures (Abou-Shleel, 2014). Extreme temperatures are responsible for a drastic change in reproductive phase and may lead to either early or late-flowering, damage to male and female reproductive tissues, flower and pod abortion (Figs. **2b** - **2d**) (Young *et al.*, 2004; Zinn *et al.*, 2010; Firon *et al.*, 2012; Djanaguiraman *et al.*, 2013). Increase in temperature also alters anther as well as pollen morphology, decreases pollen content, their dehiscence or may result in complete male sterility (Djanaguiraman *et al.*, 2013). This alteration in the course of anther dehiscence and pollen release is usually contributed by high relative humidity during heat stress (Jiang *et al.*, 2015). In comparison to female gametophytic tissue, high temperature has more devastating impacts on male gametophytic development that further affects pollen germination, viability and pollen tube elongation (Jiang *et al.*, 2012). Among all stages of pollen development, formation of meiocytes and microspores showed higher heat sensitivity, as observed in different crop species (Monterroso and Wien, 1990; Ahmed *et al.*, 1992; Djanaguiraman *et al.*, 2013). These abnormalities in anther and pollen development consequently disrupt pollination and fertilization events, hence decrease fruit and pod set drastically (Fig. **1**) (Bita and Gerats, 2013). The findings of Rainey and Griffiths (2005) showed abscission of reproductive structures as principal determinants of yield in various annual grain legumes under heat stress (Figs. **2a** - **2c**). In India, terminal heat stress is a common problem of mungbean, mostly in spring/summer season (Hanumantha Rao *et al.*, 2016). During early growth phase, high temperature *i.e.*, >40°C causes a severe loss in yield potential due to impaired fertilization, pollen sterility and high rate of flower shedding (Hanumantha Rao *et al.,* 2016). Evaluation of pollen thermotolerance and heat stress response is a concerning issue for plant geneticists, agronomists and biologists targeting improvement of the existing germplasm (Devasirvatham *et al.,* 2012; Mittler *et al.,* 2012).

In mungbean, high temperature (>40°C) has a direct effect on flower maintenance and pod development, causing up to 79% of flower shedding (Kumari and Varma 1983). In a study, 77 mutants obtained from NM 92, and 51 recombinants derived from 3 crosses *viz.,* VC1482C × NM92, VC1560D × NM92, and NM98 × VC3902A were assessed for flower retention under heat stress (Khattak *et al.,* 2006). No genotype exhibited complete tolerance to flower shedding, whereas NM 92 was sensitive to the same trait under heat stress (>40°C) (Khattak *et al.,* 2006). Further, shedding was observed only in opened flowers and not in pods at any developmental stage and fluctuations in humidity showed no effects on flower shedding (Khattak *et al.,* 2006). Like-wise, deleterious effects of heat stress (40/25°C) on two mungbean varieties (SML 832 and SML 668) were assessed, especially at reproductive stage (Kaur *et al.,* 2015). The findings revealed that high temperature *i.e* >35/25°C, 43/30°C and 45/32°C (day/night) was extremely harmful to reproductive functions, and affected potential yield of crop drastically particularly exceeding 43/30°C (Kaur *et al.,* 2015). High temperature during early growth stages and reproduction in mungbean influences seed yield adversely, due to pollen inviability, impaired fertilization, and complete flower shedding (Kaur *et al.,* 2015). Therefore, screening and selection of mungbean genotypes, which can tolerate high temperature during reproductive stages, are essential to increase its growth and productivity (Singh and Singh, 2011). Recently, a study was conducted to evaluate the effects of heat stress on 41mungbean genotypes under managed growth conditions for their vegetative and reproductive functions (Sharma *et al.,* 2016). Few selective heat tolerant mungbean lines were identified which can further contribute to future breeding programmes (Sharma *et al.,* 2016). A field trial was done by Chikukura *et al.,* (2017) consecutively in the year 2014 and 2015 during the 'kharif' season to evaluate the effects of high temperature on seven mungbean (*Vigna radiata* L. Wilczek) genotypes in rain-fed environment. Their reports showed that high temperatures *i.e.,* >44/34 °C (elevated by polyethylene sheets) significantly reduced pod number per plant as well as the seed weight/plant of mungbean (Chikukura *et al.,* 2017). Yield was reported to be highly sensitive to high temperature, and lowest yield was obtained from mungbean plants stressed at reproductive stage in both the growing years (Chikukura *et al.,* 2017). Another study investigated the variability among 28 mungbean genotypes towards their response to heat stress (45/30 °C NAc-HT, non-acclimated) particularly at reproductive stage, and to pre-acclimation of different genotypes to elevated temperatures at 35/28 °C (Ac-HT, acclimated) before exposing them to high temperatures. The total number of pollen was reduced significantly from 88/mm$^2$ in CON (28 °C/24 °C, control) to 50/mm$^2$ in Ac-HT and 40/mm$^2$ in NAc-HT plants, with visible genotypic variation, suggesting acclimated plants (Ac-HT) maintained higher pollen number and viability than non-acclimated ones (NAc-HT) (Patriyawaty *et al.,* 2018). Alagupalamuthirsolai *et al.,* (2015) also studied reproductive heat stress in 20 high yielding mungbean cultivars based on stress indices, yield contributing traits, and growing degree days (°Cd), and all the genotypes showed considerable variation for yield traits under heat stress.

High temperature interactions with flowering in some mungbean genotypes facing long photoperiods and high mean temperatures (24–28°C) have been studied thoroughly by Rawson and Craven (1979). To further increase the productivity of mungbean under heat stress environment, it is crucial to find out the genetic variation for heat tolerance in the core germplasm and probe the mechanisms governing heat sensitivity in this crop.

## Physiological And Biochemical Effects

Physiological impacts of heat stress in plants have been extensively summarized in various reports (Wang *et al.,* 2015). Generally, a mild rise in atmospheric temperature accelerates plant growth and development as well as shortens life span, subsequently resulting in marked inhibition of light assimilation throughout the plant's growth period (Kalaji *et al.,* 2016). Furthermore, interruption of elementary processes *viz.* CO$_2$ fixation, respiration, and transpiration may affect vegetative growth, metabolism and ultimately yield potential (Fig. **1**) (Maestri *et al.,* 2002). In mungbean, heat stress at vegetative stage lowers leaf photosynthesis and carbon dioxide assimilation rates, significantly, due to reduction in leaf area and size of stomatal apertures (Hanumantha Rao *et al.,* 2016) (Fig. **1**). Seven mungbean [*Vigna radiata* (L.) Wilczek] cultivars were tested for their physiological heat sensitivity by regulating changes in photosynthetic pigments, gaseous exchange attributes, malondialdehyde (MDA) and hydrogen peroxide (H$_2$O$_2$) contents in the trifoliate leaf proximal to pods (Hanif and Wahid, 2018). Further, nutrient partitioning and transport to the pods and developing seed were also measured *via* imposing heat stress at the flowering stage (Hanif and Wahid, 2018). Heat stress caused a marked accumulation of H$_2$O$_2$ and MDA, however there was a significant reduction in net photosynthesis, water use efficiency, stomatal conductance, total chlorophyll and nutrient partitioning in sensitive genotypes as compared to tolerant genotypes (Hanif and Wahid, 2018) (Fig. **1**) Like-wise, high temperatures (>45°C/35°C) affected membrane integrity (Dias *et al.,* 2010) and inhibited photosynthetic efficiency in mungbean (Kumar *et al.,* 2011). Bansal *et al.,* (2014) reported that heat stress also affects the nitrogen fixing ability of mungbean by limiting the formation of root hair and infection threads. Increase in $CO_2$ content also induces stomatal closure, hence inhibits photosynthesis; high temperature interaction studies with $CO_2$ showed negative effects on mungbean growth (Reardon and Qaderi, 2017). Findings also indicated the production of leaf starch at high temperature x $CO_2$ that results in poor assimilates transport from source to sink, hence affected grain filling adversely (Reardon and Qaderi, 2017). In another study, $CO_2$ assimilation was reduced markedly in mungbean (*Vigna radiata* L.) at 40°C, which directly affected photosynthetic efficiency (Karim *et al.,* 2003). Similarly, high temperature (>35°C) decreased chlorophyll and carotenoid contents, chlorophyll stability index and yield in different mungbean genotypes(Chand *et al.,* 2018). Anthers developing under high temperature showed diminished cell proliferation, distended vacuoles and various mitochondrial abnormalities (Sakata *et al.,* 2010). Heat stress arrests the accumulation of carbohydrates in pollen grains and stigmatic tissue by changing the assimilate partitioning and the ratio between apoplastic and symplastic phloem transport, which furthers affects pollen viability (Taiz and Zeiger, 2006; Kaur *et al.,* 2015). High temperature lowers the activity of many cell wall and vacuolar invertases, and sucrose synthase in developing pollen grains, consequently the turnover of starch and sucrose was impaired that leads to reduced accumulation of soluble carbohydrates (Kaur *et al.,* 2015). Under heat stress, RuBisCO is functional, however RuBisCO activase (RA) showed catalytic breakdown, which may stimulate destruction of total turnover activity of the enzyme (Ray *et al.,* 2003). In summer-sown mungbean genotypes, temperature exceeding 42°C caused seed hardening due to inappropriate sink development (Kaur *et al.,* 2015). Effects of heat stress on lipid peroxidation and antioxidant enzyme in four mungbean genotypes (NM 19-19, NM 20-21, NM 121-123, NM 89) revealed decreased activity of antioxidant enzymes(Mansoor and Naqvi, 2013). Further reports showed acceleration in leaf damage due to oxidative stress and poor anti-oxidative defense mechanism (Kumar *et al.,* 2013). Islam (2015) studied the effect of high temperature (36°C) on photosynthesis, leaf conductance, transpiration and yield of 8 mungbean genotypes. The study revealed that 36°C temperature at pre-flowering stage lowered leaf conductance, as compared to flowering and grain filling stages, whereas transpiration rate was not affected by high-temperature treatments at any stage, however, photosynthetic activity showed decline at all the three stages (Islam, 2015). High temperature induced reduction in chlorophyll and carotenoid contents, chlorophyll stability index and yield was also noticed in three mungbean genotypes *viz.,* MH 421, MH 318 and Basanti by Chand *et al.,* (2018). Sensitive genotypes (MH 318 and Basanti) showed more reductions in above mentioned physiological traits, while tolerant genotype (MH 421) maintained high yield and physiological functioning under heat stress (Chand *et al.,* 2018). From above it is clear that, heat stress in mungbean affects photosynthetic functions severely *via* destructing photosynthetic machinery, causing structural aberrations and alterations of chloroplast enzymes (particularly the thylakoid membrane).

## GENOTYPIC VARIANCE AND HEAT TOLERANCE

Evaluation of mungbean genotypes for genetic variation under high temperature in field condition has been reported. For instance, Khattak *et al.,* (2006) screened 14 mungbean varieties and 24 advanced cultivars selected from mutants developed by irradiation and hybridization of NM 92 with VC 1560D, VC 1482C and NM 98 x VC 3902A under high temperature for flower shedding. Lines with low flower shedding and high-pod setting under high temperature were selected in field conditions *via* marking only single plants in every segregating generation. All advanced genotypes and commercial varieties showed significant variation in plant height, days to 50% flowering, days to 90% pods maturity, 1000 seed weight and seed yield/plant under heat stress (Khattak *et al.,* 2006). The physiological and molecular aspects of the heat-resilient line EC398889 varied considerably in comparison to heat-susceptible line LGG 460, when pollen germination was examined at 44°C with the marker CEDG147 (Pratap *et al.,* 2015). An exotic line EC398889 showed high thermostability in comparison to LGG460 and molecular analysis showed significant variation among both lines for a specific marker that induces high rate of *in vivo* pollen germination under high temperature stress *i.e.,* 44 °C for 2 h (Pratap *et al.,* 2015). One of the effective means to develop heat tolerance in mungbean is to shorten the crop which may help in escaping terminal and acute heat stress during summer season. Further, this may ease synchronous podding and harvesting, faster grain filling and harmonize traits like deep root system and osmotic adjustment to evade periodic drought at vegetative stages and terminal heat >40 °C during seed filling (Naveed *et al.,* 2015). Considering multi-location heat stress examination at Durgapura (Rajasthan) and Vamban (Tamilnadu) in India, 12 novel genotypes (IPM 02-16, IPM 9901-10, IPM 409-4, IPM 02-3, PDM 139, IPM 02-1, IPM 2-14, IPM 9-43-K, PDM 288, EC 470096, IPM 2K14-9, IPM 2K14-5) have been identified with high tolerance to heat as well as drought (Basu *et al.,* 2009). In another study, a few heat-tolerant mungbean varieties were identified on the basis of protein profiling and sucrose synthase activity as biochemical markers, and were further validated by multiple field trials across different agro-climatic zones sensitive to intermittent heat stress. These genotypes are PDM 139 (Samrat), IPM 02-1, PDM 288, IPM 05-3-21, ML-1257. In another heat stress experiment, three cultivars *viz.,* IPM 02-3, IPM 2-14, Samrat, were marked as the heat tolerant on the basis of their low heat susceptibility index (Alagupalamuthirsolai *et al.,* 2015). Some short-duration mungbean genotypes (60-70 days maturity time) have been developed recently for introducing them between the rabi-kharif cropping cycles (Kaur *et al.,* 2015; Sharma *et al.,* 2016). These genotypes are being tested for their response to heat stress, which is likely to occur during their reproductive stage.

From above it is clear that to sustain mungbean productivity under climate change scenario, deployment of heat tolerant mungbean varieties, which are developed either through conventional or modern breeding, is necessary (Patriyawaty *et al.,* 2018). By exploring molecular approaches such as transcriptomic and metabolomics, researchers can breed high yielding, durable genotypes, which can withstand different types of stresses in changing the environment as suggested by Hanumantha Rao *et al.,* (2016). Evidently, no molecular marker-assisted breeding programmes related to heat tolerance in mungbean are elucidated till now, hence appropriate depiction of gene/s function and action against heat stress is still illusive and need to be explored (Hanumantha Rao *et al.,* 2016). As data regarding physiological mechanisms for genotypic variation under high temperature stress in mungbean (*Vigna radiata*) is rarely available, identification of traits that contribute in plant's acclimation to heat stress environments is necessary (Patriyawaty *et al.,* 2018).

## AGRONOMIC PRACTICES FOR HEAT RESILIENCE

Crop yield and productivity usually depends on the interaction between environment and genotype's potential expression, which can be manipulated by different agronomic managements (Zandalinas *et al.,* 2018). On the basis of adaptation ability, every genotype has varying yield capabilities (Vinocur & Altman, 2005). Plant cultivation and agricultural systems are evolving continuously due to introduction of new agronomic tools and development of high-yielding and stress tolerant cultivars *via* genetic improvement (Mariani and Ferrante, 2017). Although, it's impossible to influence temperatures during crop growth in an open field, few management practices can be implemented with some success. The most common method is to adjust the date of sowing to avoid damaging heat effects during late growth periods. Optimization of agronomic practices such as irrigation management, methods of fertilization, appropriate sowing time and the use of exogenous protectants may alleviate devastating impacts of heat stress in mungbean.

### Soil and Water Management

Alteration in upper layer of soil affects water and heat balance in terms of increase in evaporation, infiltration and thermal exchange between soil and atmosphere (Ferrero *et al.,* 2005; Yadav *et al.,* 2010). One important method to limit water evaporation under heat stress is superficial tillage, surface residues and soil mulching (Fig. **3**) (Ferrero *et al.,* 2005). Another agronomic strategy to minimize the harmful effects of high temperature is to maintain high soil water content to improve thermo-regulation, if adequate quantity of water is available (Ferrero *et al.,* 2005). soil surface texture, temperature and water vapor gradients, and infiltration processes determine the amount of water stored by the soil and uptake by plants (Lipiec *et al.,* 2006; 2012). Further, application of organic mulch on soil surface lowers the soil temperature attributed to its low thermal conductivity (Khan *et al.,* 2000) and maintains water level by monitoring surface evaporation (Mulumba and Lal, 2008). Deep tilling also helps in increasing root growth in soils with definite hard sub-soils, however, it is practiced only in dense soil paddocks due to its high cost (Siczek and Lipiec 2011; Martínez *et al.,* 2012).Current irrigation schemes to escape sunburn includes drip and film hole irrigation and sprinkling that save up to >50% water and improve water use efficiency (WUE) and crop yield compared to overhead irrigation (Yadav *et al.,* 2010).

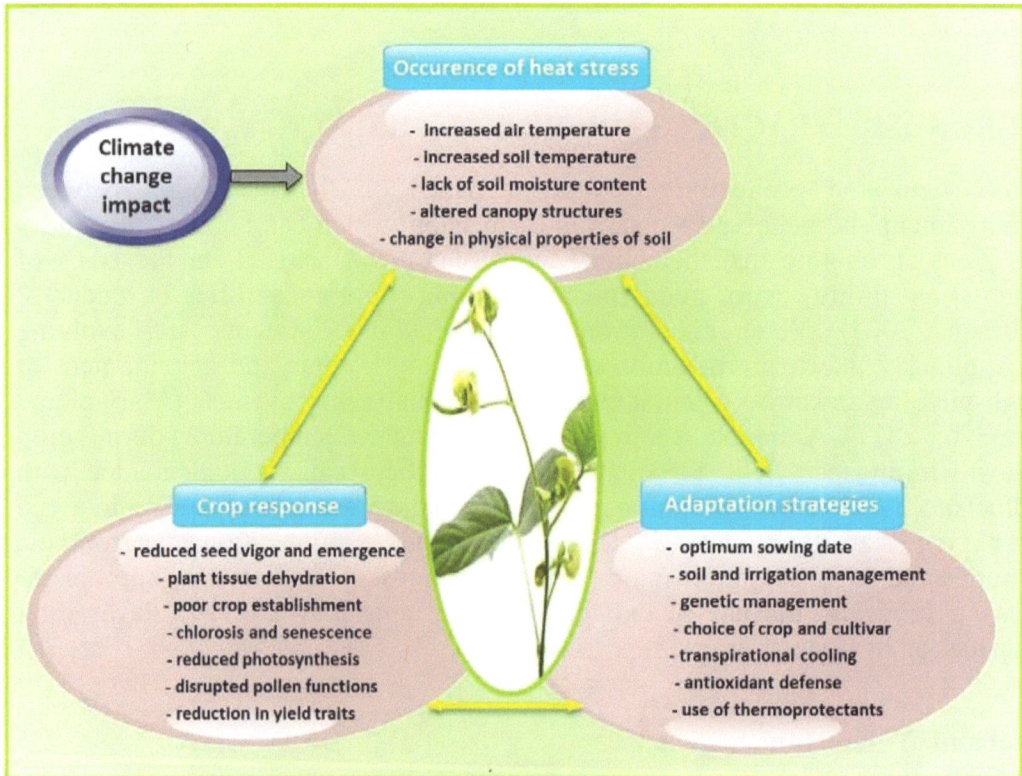

**Fig. (3).** Relationship between heat stress due to climate change, crop responses to heat stress, and different adaptation strategies in the 'field conditions.

## Optimum Sowing Time And Nutrient Management

Crops vary in their response to high temperature and drought conditions and inherent genetic factors help in withstanding adversities of these stresses (Blum, 2005; Singh and Singh, 2011). Generally, selection of crop types and varieties determine crop yield and productivity under different stress factors (Fig. **3**). Varieties that mature early show better performance in heat as well as drought prone areas *via* reducing terminal heat and soil evaporation due to early development of crop stand and canopy structures (Yadav *et al.,* 2010). Generally, the late sown mungbean, particularly during summer season, faces severe temperature stress that shortens the grain filling and maturity duration, consequently affecting final yield and grain quality (Hanumantha Rao *et al.,* 2016). Therefore, to escape terminal heat stress and to develop high yielding mungbean cultivars adapted to arid and semiarid regions, genotypes with early maturity and appropriate grain-filling time should be selected. Hence, maintenance of proper sowing time may act as significant agronomic practice for attaining optimum plant growth and yield under heat-stress environment (Kajla *et al.,* 2015).

The optimum use of plant nutrients and their management may play crucial role in developing plant resilience to high temperature stress. Waraich *et al.,* (2012) suggested that appropriate plant nutrition is effective in mitigating detrimental effects of heat stress *via* various means. Heat stress increases reactive oxygen species (ROS) production due in stressed plants, which in turns causes photo-oxidation and destructs chloroplast membranes (Fig. **1**). However, additional supply of plant nutrients reduces ROS toxicity *via* stimulating antioxidants production in plant cells. These accumulated antioxidants further help in ROS detoxification, maintenance of membrane integrity and enhance photosynthesis. The added plant nutrients are also helpful in maintaining high tissue water potential during heat stress condition.

## Genetic Management

Genetic resources of mungbean are maintained throughout the globe at different research centers including The World Vegetable Center, Taiwan; The University of the Philippines; National Bureau of Plant Genetic Resources of the Indian Council of Agricultural Research; Indian Institute of pulses Research, Kanpur; The Plant Genetic Resources Conservation Unit of the University of Georgia, USA; and The Institute of Crop Germplasm Resources of the Chinese Academy of Agricultural Sciences (Ebert, 2013). Additionally the Rural Development Administration (RDA), Korea, and the University of The Philippines also have duplicate collection of some parts of the mungbean germplasm maintained at The World Vegetable Center. To facilitate efficient utilization of genetic resources and to provide access to mungbean scientists, core germplasm collections have been established in countries like India, China, Korea and the USA. Recently, a core collection with 1481 accessions and a mini-core with 296 accessions have been developed by AVRDC-The World Vegetable Center (Schafleitner *et al.,* 2015) .Generally, the core collection was prepared considering different phenotypic characters, while the mini-core was prepared by using 20 SSR markers *via* molecular characterization (Schafleitner *et al.,* 2015). This minicore set is being effectively deployed for trait introgression in elite mungbean backgrounds under the aegis of 'International mungbean improvement network' in which IIPR, Kanpur; BARI, Bangladesh; DAR Myanmar, World Vegetable Centre, Taiwan and Hyderabad and University of Queensland, Australia are the partners. A few more partners have been added recently in the network (Nair *et al.,* 2019).

### Screening Germplasm For Heat Stress Tolerance

As breeding is a vital strategy to develop crops under changing climate conditions, the evaluation of genetic diversity and selection and introduction of stress resistance genes for developing new varieties is highly recommended in the agricultural production systems (Chapman *et al.,* 2012). Breeding strategies for heat stress resilience are required for genetic improvement of mungbean to heat stress. Recently, various studies have been done to find out heat tolerant mungbean genotypes (Kaur *et al.,* 2015; Sharma *et al.,* 2016) by using various screening techniques (Fig. **3**) (Sehrawat *et al.,* 2013; Basu *et al.,* 2019). In breeding for heat-tolerance, a variety of physiological approaches have been found to be effective and the methodology generally includes screening genetic resources for identification of donors for heat tolerance in crops. Desired new plant varieties can be developed through hybridization and selection for novel traits, to overcome future climatic aberrations.

### Biotechnological/Transgenic Approach For Improving Heat Tolerance

Genetic engineering can help in alleviating the harmful effects of high temperature stress by developing heat tolerant genotypes (Chapman *et al.,* 2012). The process involves the introgression of desired gene of interest into the target plants to impart thermo-tolerance (Zheng *et al.,* 2012). As mungbean has small genome size, it can act as a model plant among different *Vigna* species, however, due to complex genomic patterns and high recalcitrancy, genetic modification in mungbean is highly difficult (Kim *et al.,* 2015). Hence, a thorough insight of the regulatory mechanisms underlying heat stress or other environmental responses is necessary to develop genotypes that can overcome changing climate patterns and are able to grow in harsh environments particularly in arid and semi-arid tropics (Kim *et al.,* 2015). By collecting and re-sequencing wild mungbean lines from different geographic regions, researchers may find allelic variation for beneficial traits mined from wild mungbean that may further help in expansion of mungbean worldwide (Kim *et al.,* 2015). Interestingly, a wild genome sequence of mungbean was released by Kang *et al.,* (2014) and proper utilization of this genomic resource may act as promising route for translational genomics studies for heat stress tolerance.

Recently, various transcription factors (TFs) associated with heat stress tolerance have been found and engineered successfully to improve stress tolerance in crops (Wang *et al.,* 2016). These transcription factors can be utilized for improving thermo-period tolerance in mungbean too; however no such studies have been carried out so far in mungbean. These advancements in biotechnological approaches can bring large scale structural modifications that are known to play a crucial role in the adaptation of the mungbean crop to heat stress environments along with other stresses.

## Exogenous Application of Thermo-Protectants

The increased application of some chemical protectants in the form of phytohormones (brassiosteroids, SA, ABA), signaling molecules, trace elements (selenium *etc.*) and osmoprotectants (proline, glycine betaine, trehalose) have proven to be highly beneficial in plants facing heat stress as these molecules possess antioxidant and growth-promoting abilities (Fig. **3**) (Akhtar *et al.,* 2015). As heat-susceptible cultivars are apparently unable to accumulate these substances, exogenous application of these protective molecules may impart heat tolerance in such plants (Rasheed *et al.,* 2011). These thermo-protectants play promising role in protecting plants from harmful impacts of high-temperature stress and provide defense by detoxifying ROS *via* up-regulation of antioxidant mechanisms (Awasthi *et al.,* 2015).

Heat stress (50°C) at early developmental stages show marked reduction in amylase synthesis, however, application of gibberellic acid (100 μM) results in high amylase induction, improving the growth of mungbean seedlings as studied by Mansoor and Naqvi (2013). It has also been observed that heat shock treatments with salicylic acid (SA) (0.5 or 1mM) application helps in alleviating deteriorative heat symptoms such as lipid peroxidation, electrolyte leakage, and increased level of $H_2O_2$ in mungbean seedlings, hence contributing to enhanced activity of the free-radical scavenging systems (Saleh *et al.,* 2007). The study reported that exogenous application of salicylic acid (SA, 0.5 or 1mM) after 15 days of seeding helps in decreasing deteriorative impacts of increased temperature such as lipid peroxidation, electrolyte leakage, and increased $H_2O_2$ level in mungbean seedlings and also contribute to the enhanced activity of the free-radical scavenging systems (Saleh *et al.,* 2007). Exogenous application of ascorbic acid (ASC, 50 μM) at the time of seedling growth also conferred thermotolerance in heat-stressed mungbean plants as reported by Kumar *et al.,* 2011. Nahar *et al.,* 2015 reported the harmful effects of heat stress (42°C) in mungbean and found that high temperature for 48 h results in oxidative damage and increases MG (Methyl glyoxal) toxicity, which can be mitigated by exogenous application of glutathione. In a recent study, exogenous proline treatment (5mM) in heat-stressed mungbean plants increased its endogenous levels in both vegetative as well reproductive organs, which improved pollen fertility, stigma receptivity and ovule functions, significantly (Priya *et al.,* 2019). In addition, stress damage to leaves decreased and chlorophyll content increased with proline treatment that ultimately improved carbon fixation and assimilation to increase the flower formation, pod setting, filled pods, and pod and seed weight per plants, suggesting its crucial role in imparting thermo-tolerance (Priya *et al.,* 2019). Another study showed that the treatment of heat-stressed mungbean plants with GABA (γ-aminobutyric acid; a non-protein amino acid; 1mM) improved pollen viability, pollen germination, stigma receptivity and ovule viability significantly compared to untreated HS controls (Priya *et al.,* 2019). Further, GABA-treated heat-stressed plants showed less damage to membranes, carbon assimilation rates (sucrose synthesis and its utilization) and photosynthetic machinery (chlorophyll content, chlorophyll fluorescence, RuBisCO activity) (Priya *et al.,* 2019). The positive response of these protective molecules under heat stress is possibly associated with reduction in oxidative damage and increased activity of various antioxidants. These findings clearly showed that such osmoprotectants have improved the growth of heat-stressed mungbean plants significantly, thus may prove as potential regulators in engineering thermotolerance in plants.

## CONCLUSIONS

High temperature stress causes tremendous losses to mungbean productivity worldwide. Although, intensive work has been carried out to assess the deleterious effects of heat stress in mungbean, the understanding of various underlying tolerance mechanisms still remains elusive. To develop high yielding and heat-resilient cultivars, systematic knowledge of various metabolic and developmental processes along with regulative mechanisms associated with the stress is necessary. While, a noticeable progression in understanding the heat stress response in mungbean has been achieved, further insight regarding various biochemical and molecular means of heat tolerance is required to improve the crop yield under challenging warmer environments. Under extreme temperatures, the functional genomics tools would be supportive in sustaining mungbean yield. Adoption of improved management practices and exploitation of genetic and genomic tools for developing climate resilient cultivars may help in overcoming the complexities of heat stress. The exogenous applications of thermo-protectants have proved beneficial in improving heat tolerance in mungbean under controlled environment studies, which need to be further validated under realistic field environments. Although some level of heat tolerance has been found in mungbean, complete tolerance has not been achieved yet, and hence, an integrated effort of plant physiologists, molecular biologists and mungbean breeders is required. Further, systematic analysis of mungbean genome is required to ensure an accurate mapping of different traits, introduction of resistant alleles, or cloning of associated QTLs for heat tolerance, which may facilitate identification of genes involved in heat tolerance. Marker-assisted breeding programs must be promoted for introgression of heat stress-resistant genes and QTLs to attain effective heat tolerance in mungbean.

## CONSENT FOR PUBLICATION

Not applicable.

## CONFLICT OF INTEREST

The authors confirm that this chapter content has no conflict of interest.

## ACKNOWLEDGEMENTS

Declared none.

## REFERENCES

Abdel lateff, E. (2018). Effect of climate change on mungbean growth and productivity under Egyptian conditions. *Int. J. Agricult. Forest Life Sci., 2*(2), 16-23.

Abou-Shleel, S.M. (2014). Effect of air temperature on growth, yield and active ingredients of fenugreek (Trigonella foenum-graecum). *Nat. Sci., 12*, 50-54.

Ahmed, F.E., Hall, A.E., DeMason, D.A. (1992). Heat injury during floral development in cowpea (*Vigna unguiculata, Fabaceae*). *Am. J. Bot., 79*(7), 784-791.
[http://dx.doi.org/10.1002/j.1537-2197.1992.tb13655.x]

Akhtar, S., Naik, A., Hazra, P. (2015). Harnessing heat stress in vegetable crops towards mitigating impacts of climate change. *Clim Dynamic Horticult. Sci., 173.*

Alagupalamuthirsolai, M., Vijaylakshmi, C., Basu, P.S., Singh, J. (2015). Physiological evaluation of mungbean (*Vigna radiata (l.)* Wilczek) cultivars for heat tolerance. *Indian J. Pulses Res., 28*(2), 30-34.

Awasthi, R., Bhandari, K., Nayyar, H. (2015). Temperature stress and redox homeostasis in agricultural crops. *Front. Environ. Sci., 3*, 11.
[http://dx.doi.org/10.3389/fenvs.2015.00011]

Bansal, M., Kukreja, K., Sunita, S., Dudeja, S.S. (2014). Symbiotic effectivity of high temperature tolerant mungbean (*Vigna radiata*) rhizobia under different temperature conditions. *Int. J. Curr. Microbiol. Appl. Sci., 3*(12), 807-821.

Basu, P.S., Ali, M., Chaturvedi, S.K. Terminal heat stress adversely affects chickpea productivity in Northern India–strategies to improve thermotolerance in the crop under climate change. *In 'ISPRS Archieves XXXVIII-8/W3 Workshop Proceedings: Impact of Climate Change on Agriculture'*. 23-25 Feb. 2009 New Delhi, India 189-193.

Basu, P.S., Pratap, A., Gupta, S., Sharma, K., Tomar, R., Singh, N.P. (2019). Physiological traits for shortening crop duration and improving productivity of greengram (*Vigna radiata L.* Wilczek) under high temperature. *Front. Plant Sci., 10*, 1508.
[http://dx.doi.org/10.3389/fpls.2019.01508] [PMID: 31867025]

Biswash, M.R., Rahman, M.W., Haque, M.M., Sharmin, M., Barua, R. (2014). Effect of potassium and vermicompost on the growth, yield and nutrient contents of mungbean (BARI Mung 5). *Open Sci. J. Biosci. Bioeng., 1*(3), 33-39.

Bita, C.E., Gerats, T. (2013). Plant tolerance to high temperature in a changing environment: scientific fundamentals and production of heat stress-tolerant crops. *Front. Plant Sci., 4*, 273.
[http://dx.doi.org/10.3389/fpls.2013.00273] [PMID: 23914193]

Blum, A. (1989). Breeding methods for drought resistance. In: H. G., Jones, T. J., Flowers, M. B., Jones, (Eds.), *Plant Under Stress,* Cambridge University Press.

Blum, A. (2005). Drought resistance, water-use efficiency, and yield potential—are they compatible, dissonant, or mutually exclusive? *Aust. J. Agric. Res., 56*(11), 1159-1168.
[http://dx.doi.org/10.1071/AR05069]

Cao, D., Li, H., Yi, J., Zhang, J., Che, H., Cao, J., Yang, L., Zhu, C., Jiang, W. (2011). Antioxidant properties of the mung bean flavonoids on alleviating heat stress. *PLoS One, 6*(6), e21071.
[http://dx.doi.org/10.1371/journal.pone.0021071] [PMID: 21695166]

Chand, G., Nandwal, A.S., Kumar, N., Devi, S., Khajuria, S. (2018). Yield and physiological responses of mungbean Vigna radita (L.) Wilczek genotypes to high temperature at reproductive stage. *Legume Research-An International Journal, 41*(4), 557-562.
[http://dx.doi.org/10.18805/LR-3795]

Chapman, S.C., Chakraborty, S., Dreccer, M.F., Howden, S.M. (2012). Plant adaptation to climate change—opportunities and priorities in breeding. *Crop Pasture Sci., 63*(3), 251-268.
[http://dx.doi.org/10.1071/CP11303]

Chauhan, Y.S., Williams, R. (2018). Physiological and agronomic strategies to increase Mungbean yield in climatically variable environments of Northern Australia. *Agronomy (Basel), 8*(6), 83.
[http://dx.doi.org/10.3390/agronomy8060083]

Chikukura, L., Bandyopadhyay, S.K., Kumar, S.N., Pathak, H., Chakrabarti, B. (2017). Effect of elevated temperature stress on growth, yield and yield attributes of mungbean (*Vigna radiata*) in semi-arid north-west India. *Curr. Adv. Agric. Sci., 9*(1), 18-22. [An International Journal].
[http://dx.doi.org/10.5958/2394-4471.2017.00003.X]

Dahiya, P.K., Linnemann, A.R., Van Boekel, M.A.J.S., Khetarpaul, N., Grewal, R.B., Nout, M.J.R. (2015). Mung bean: technological and nutritional potential. *Crit. Rev. Food Sci. Nutr., 55*(5), 670-688.
[http://dx.doi.org/10.1080/10408398.2012.671202] [PMID: 24915360]

Devasirvatham, V., Gaur, P.M., Mallikarjuna, N., Tokachichu, R.N., Trethowan, R.M., Tan, D.K.Y. (2012). Effect of high temperature on the reproductive development of chickpea genotypes under controlled environments. *Funct. Plant Biol., 39*(12), 1009-1018.
[http://dx.doi.org/10.1071/FP12033] [PMID: 32480850]

Dias, A.S., Barreiro, M.G., Campos, P.S., Ramalho, J.C., Lidon, F.C. (2010). Wheat cellular membrane thermotolerance under heat stress. *J. Agron. Crop Sci., 196*(2), 100-108.

[http://dx.doi.org/10.1111/j.1439-037X.2009.00398.x]

Dickson, M.H., Boettger, M.A. (1984). Effect of high and low temperatures on pollen germination and seed set in snap beans. *J. Am. Soc. Hortic. Sci., 109*(3), 372-374.

Djanaguiraman, M., Prasad, P.V., Boyle, D.L., Schapaugh, W.T. (2013). Soybean pollen anatomy, viability and pod set under high temperature stress. *J. Agron. Crop Sci., 199*(3), 171-177.
[http://dx.doi.org/10.1111/jac.12005]

Ebert, A. W. (2013). Sprouts, microgreens, and edible flowers: The potential for high value specialty produce in Asia. *SEAVEG 2012: High Value Vegetables in Southeast Asia: Production, Supply and Demand*, 216-227.

Farooq, M., Bramley, H., Palta, J.A., Siddique, K.H. (2011). Heat stress in wheat during reproductive and grain-filling phases. *Crit. Rev. Plant Sci., 30*(6), 491-507.
[http://dx.doi.org/10.1080/07352689.2011.615687]

Farooq, M., Hussain, M., Usman, M., Farooq, S., Alghamdi, S.S., Siddique, K.H.M. (2018). Impact of abiotic stresses on grain composition and quality in food legumes. *J. Agric. Food Chem., 66*(34), 8887-8897.
[http://dx.doi.org/10.1021/acs.jafc.8b02924] [PMID: 30075073]

Fedoroff, N. V., Battisti, D. S., Beachy, R. N., Cooper, P. J., Fischhoff, D. A., Hodges, C. N., Reynolds, M. P. (2010). Radically rethinking agriculture for the 21ˢᵗ century. *Science, 327*(5967), 833-834.

Ferrero, A., Usowicz, B., Lipiec, J. (2005). Effects of tractor traffic on spatial variability of soil strength and water content in grass covered and cultivated sloping vineyard. *Soil Tillage Res., 84*(2), 127-138.
[http://dx.doi.org/10.1016/j.still.2004.10.003]

Firon, N., Nepi, M., Pacini, E. (2012). Water status and associated processes mark critical stages in pollen development and functioning. *Ann. Bot., 109*(7), 1201-1214.
[http://dx.doi.org/10.1093/aob/mcs070] [PMID: 22523424]

Godfray, H. C. J., Beddington, J. R., Crute, I. R., Haddad, L., Lawrence, D., Muir, J. F., Toulmin, C. (2010). Food security: The challenge of feeding 9 billion people. *Science, 327*(5967), 812-818.

Gounaris, K., Brain, A.R.R., Quinn, P.J., Williams, W.P. (1984). Structural reorganisation of chloroplast thylakoid membranes in response to heat-stress. *Biochimica et Biophysica Acta (BBA)-. Bioenergetics, 766*(1), 198-208.
[http://dx.doi.org/10.1016/0005-2728(84)90232-9]

Guo, M., Liu, J.H., Ma, X., Luo, D.X., Gong, Z.H., Lu, M.H. (2016). The plant heat stress transcription factors (HSFs): structure, regulation, and function in response to abiotic stresses. *Front. Plant Sci., 7*, 114.
[http://dx.doi.org/10.3389/fpls.2016.00114] [PMID: 26904076]

Hall, A.E. (1992). Breeding for heat tolerance. *Plant Breed. Rev., 10*(2), 129-168.

Hamada, A.M. (2001). Alteration in growth and some relevant metabolic processes of broad bean plants during extreme temperatures exposure. *Acta Physiol. Plant., 23*(2), 193-200.
[http://dx.doi.org/10.1007/s11738-001-0008-y]

Hanif, A., Wahid, A. (2018). Seed yield loss in mungbean is associated to heat stress induced oxidative damage and loss of photosynthetic capacity in proximal trifoliate leaf. *Pak. J. Agric. Sci., 55*(4)

Hanumantha Rao, B., Nair, R.M., Nayyar, H. (2016). Salinity and high temperature tolerance in mungbean [*Vigna radiata (L.)* Wilczek] from a physiological perspective. *Front. Plant Sci., 7*, 957.
[http://dx.doi.org/10.3389/fpls.2016.00957] [PMID: 27446183]

Hatfield, J.L., Prueger, J.H. (2015). Temperature extremes: Effect on plant growth and development. *Weather Clim. Extrem., 10*, 4-10.
[http://dx.doi.org/10.1016/j.wace.2015.08.001]

Hossain, A. (2012). Phenology, growth and yield of three wheat (*Triticum aestivum L.*) varieties as affected by high temperature stress. *Not. Sci. Biol., 4*(3), 97-109.

[http://dx.doi.org/10.15835/nsb437879]

Islam, M.T. (2015). Effects of high temperature on photosynthesis and yield in mungbean. *Bangladesh J. Bot., 44*(3), 451-454.
[http://dx.doi.org/10.3329/bjb.v44i3.38553]

Jha, U.C., Bohra, A., Parida, S.K., Jha, R. (2017). Integrated "omics" approaches to sustain global productivity of major grain legumes under heat stress. *Plant Breed., 136*(4), 437-459.
[http://dx.doi.org/10.1111/pbr.12489]

Jiang, Y., Lahlali, R., Karunakaran, C., Kumar, S., Davis, A.R., Bueckert, R.A. (2015). Seed set, pollen morphology and pollen surface composition response to heat stress in field pea. *Plant Cell Environ., 38*(11), 2387-2397.
[http://dx.doi.org/10.1111/pce.12589] [PMID: 26081983]

Kajla, M., Yadav, V.K., Khokhar, J., Singh, S., Chhokar, R.S., Meena, R.P., Sharma, R.K. (2015). Increase in wheat production through management of abiotic stresses: a review. *J. Appl. Nat. Sci., 7*(2), 1070-1080.
[http://dx.doi.org/10.31018/jans.v7i2.733]

Kalaji, H.M., Jajoo, A., Oukarroum, A., Brestic, M., Zivcak, M., Samborska, I.A. (2016). Chlorophyll a fluorescence as a tool to monitor physiological status of plants under abiotic stress conditions. *Acta Physiol. Plant., 38*(4), 102.
[http://dx.doi.org/10.1007/s11738-016-2113-y]

Kang, Y.J., Kim, S.K., Kim, M.Y., Lestari, P., Kim, K.H., Ha, B.K., Jun, T.H., Hwang, W.J., Lee, T., Lee, J., Shim, S., Yoon, M.Y., Jang, Y.E., Han, K.S., Taeprayoon, P., Yoon, N., Somta, P., Tanya, P., Kim, K.S., Gwag, J.G., Moon, J.K., Lee, Y.H., Park, B.S., Bombarely, A., Doyle, J.J., Jackson, S.A., Schafleitner, R., Srinives, P., Varshney, R.K., Lee, S.H. (2014). Genome sequence of mungbean and insights into evolution within Vigna species. *Nat. Commun., 5*, 5443.
[http://dx.doi.org/10.1038/ncomms6443] [PMID: 25384727]

Karim, A., Fukamachi, H., Hidaka, T. (2003). Photosynthetic performance of *Vigna radiata L.* leaves developed at different temperature and irradiance levels. *Plant Sci., 164*(4), 451-458.
[http://dx.doi.org/10.1016/S0168-9452(02)00423-5]

Kaur, R., Bains, T.S., Bindumadhava, H., Nayyar, H. (2015). Responses of mungbean (*Vigna radiata L.*) genotypes to heat stress: Effects on reproductive biology, leaf function and yield traits. *Sci. Hortic. (Amsterdam), 197*, 527-541.
[http://dx.doi.org/10.1016/j.scienta.2015.10.015]

Kaushal, N., Bhandari, K., Siddique, K.H., Nayyar, H. (2016). Food crops face rising temperatures: an overview of responses, adaptive mechanisms, and approaches to improve heat tolerance. *Cogent Food Agric., 2*(1)1134380
[http://dx.doi.org/10.1080/23311932.2015.1134380]

Keatinge, J.D.H., Easdown, W.J., Yang, R.Y., Chadha, M.L., Shanmugasundaram, S. (2011). Overcoming chronic malnutrition in a future warming world: the key importance of mungbean and vegetable soybean. *Euphytica, 180*(1), 129-141.
[http://dx.doi.org/10.1007/s10681-011-0401-6]

Khan, A.R., Chandra, D., Quraishi, S., Sinha, R.K. (2000). Soil aeration under different soil surface conditions. *J. Agron. Crop Sci., 185*(2), 105-112.
[http://dx.doi.org/10.1046/j.1439-037X.2000.00417.x]

Khattak, G.S.S., Saeed, I.Q.B.A.L., Muhammad, T. (2006). Breeding for heat tolerance in mungbean (*Vigna radiata (L.)* Wilczek). *Pak. J. Bot., 38*(5), 1539-1550.

Kim, S.K., Nair, R.M., Lee, J., Lee, S.H. (2015). Genomic resources in mungbean for future breeding programs. *Front. Plant Sci., 6*, 626.
[http://dx.doi.org/10.3389/fpls.2015.00626] [PMID: 26322067]

Kumar, S., Kaur, R., Kaur, N., Bhandhari, K., Kaushal, N., Gupta, K. (2011). Heat-stress induced inhibition in growth and chlorosis in mungbean (Phaseolus aureus Roxb.) is partly mitigated by ascorbic acid application and is related to reduction in oxidative stress. *Acta Physiol. Plant., 33*(6), 2091. [http://dx.doi.org/10.1007/s11738-011-0748-2]

Kumar, S., Thakur, P., Kaushal, N., Malik, J.A., Gaur, P., Nayyar, H. (2013). Effect of varying high temperatures during reproductive growth on reproductive function, oxidative stress and seed yield in chickpea genotypes differing in heat sensitivity. *Arch. Agron. Soil Sci., 59*(6), 823-843. [http://dx.doi.org/10.1080/03650340.2012.683424]

Kumari, P., Varma, S.K. (1983). Genotypic differences in flower production/shedding and yield in mungbean (*Vigna radiata*). *Indian J. Plant. Physiol., 26*(4), 402-405.

Kurets, V.K., Popov, E.G. (1988). Evaluating the requirements of a genotype in respect of environmental conditions. Diagnostika urtoichivosti rastenii stressovym vozdeistviyam. *Diagnostika urtoichivosti rastenii stressovym vozdeistviyam, Leningrad, USSR,* 222-227.

Lambrides, C.J., Godwin, I.D. (2007). *Mungbean, Pulses, Sugar and Tuber Crops.* (pp. 69-90). Berlin, Heidelberg: Springer. [http://dx.doi.org/10.1007/978-3-540-34516-9_4]

Lawn, R.J. (1989). Agronomic and physiological constraints to the productivity of tropical grain legumes and prospects for improvement. *Exp. Agric., 25*(4), 509-528. [http://dx.doi.org/10.1017/S0014479700015143]

Lipiec, J., Horn, R., Pietrusiewicz, J., Siczek, A. (2012). Effects of soil compaction on root elongation and anatomy of different cereal plant species. *Soil Tillage Res., 121*, 74-81. [http://dx.doi.org/10.1016/j.still.2012.01.013]

Lipiec, J., Kuś, J., Słowińska-Jurkiewicz, A., Nosalewicz, A. (2006). Soil porosity and water infiltration as influenced by tillage methods. *Soil Tillage Res., 89*(2), 210-220. [http://dx.doi.org/10.1016/j.still.2005.07.012]

Nair, R.M., Pandey, A.K., War, A.R., Hanumantha rao, B., Shwe, T., Alam, A., Pratap, A., Malik, S.R., Karimi, R., Mbeyagala, E.K., Douglas, C.A., Rane, J., Schafleitner, R. (2019). Biotic and abiotic constraints in mungbean production-progress in genetic improvement. *Front. Plant Sci., 10*, 1340. [http://dx.doi.org/10.3389/fpls.2019.01340] [PMID: 31736995]

Maestri, E., Klueva, N., Perrotta, C., Gulli, M., Nguyen, H.T., Marmiroli, N. (2002). Molecular genetics of heat tolerance and heat shock proteins in cereals. *Plant Mol. Biol., 48*(5-6), 667-681. [http://dx.doi.org/10.1023/A:1014826730024] [PMID: 11999842]

Mannur, D.M., Babbar, A., Thudi, M., Sabbavarapu, M.M., Roorkiwal, M., Yeri, S.B., Bansal, V.P., Jayalakshmi, S.K., Singh Yadav, S., Rathore, A., Chamarthi, S.K., Mallikarjuna, B.P., Gaur, P.M., Varshney, R.K. (2019). Super Annigeri 1 and improved JG 74: two Fusarium wilt-resistant introgression lines developed using marker-assisted backcrossing approach in chickpea (*Cicer arietinum* L.). *Mol. Breed., 39*(1), 2. [http://dx.doi.org/10.1007/s11032-018-0908-9] [PMID: 30631246]

Mansoor, S., Naqvi, F.N. (2013). Isoamylase profile of mung bean seedlings treated with high temperature and gibberellic acid. *Afr. J. Biotechnol., 12*(13).

Mariani, L., Ferrante, A. (2017). Agronomic management for enhancing plant tolerance to abiotic stresses—drought, salinity, hypoxia, and lodging. *Horticulturae, 3*(4), 52. [http://dx.doi.org/10.3390/horticulturae3040052]

Martínez, I.G., Prat, C., Ovalle, C., del Pozo, A., Stolpe, N., Zagal, E. (2012). Subsoiling improves conservation tillage in cereal production of severely degraded Alfisols under Mediterranean climate. *Geoderma, 189*, 10-17. [http://dx.doi.org/10.1016/j.geoderma.2012.03.025]

Mittler, R. (2006). Abiotic stress, the field environment and stress combination. *Trends Plant Sci., 11*(1), 15-19.

[http://dx.doi.org/10.1016/j.tplants.2005.11.002] [PMID: 16359910]

Mittler, R., Finka, A., Goloubinoff, P. (2012). How do plants feel the heat? *Trends Biochem. Sci., 37*(3), 118-125.
[http://dx.doi.org/10.1016/j.tibs.2011.11.007] [PMID: 22236506]

Monterroso, V.A., Wien, H.C. (1990). Flower and pod abscission due to heat stress in beans. *J. Am. Soc. Hortic. Sci., 115*(4), 631-634.
[http://dx.doi.org/10.21273/JASHS.115.4.631]

Mulumba, L.N., Lal, R. (2008). Mulching effects on selected soil physical properties. *Soil Tillage Res., 98*(1), 106-111.
[http://dx.doi.org/10.1016/j.still.2007.10.011]

Nahar, K., Hasanuzzaman, M., Alam, M.M., Fujita, M. (2015). Exogenous glutathione confers high temperature stress tolerance in mung bean (*Vigna radiata L.*) by modulating antioxidant defense and methylglyoxal detoxification system. *Environ. Exp. Bot., 112*, 44-54.
[http://dx.doi.org/10.1016/j.envexpbot.2014.12.001]

Naveed, M., Mehboob, I., Hussain, M.B., Zahir, Z.A. (2015). Perspectives of rhizobial inoculation for sustainable crop production. *Plant Microbes Symbiosis: Applied Facets.* (pp. 209-239). New Delhi: Springer.
[http://dx.doi.org/10.1007/978-81-322-2068-8_11]

Ohama, N., Sato, H., Shinozaki, K., Yamaguchi-Shinozaki, K. (2017). Transcriptional regulatory network of plant heat stress response. *Trends Plant Sci., 22*(1), 53-65.
[http://dx.doi.org/10.1016/j.tplants.2016.08.015] [PMID: 27666516]

Patriyawaty, N.R., Rachaputi, R.C., George, D., Douglas, C. (2018). Genotypic variability for tolerance to high temperature stress at reproductive phase in Mungbean. *Sci. Hortic. (Amsterdam), 227*, 132-141. [*Vigna radiata (L.) Wilczek*].
[http://dx.doi.org/10.1016/j.scienta.2017.09.017]

Piramila, B.H.M., Prabha, A.L., Nandagopalan, V., Stanley, A.L. (2012). Effect of heat treatment on germination, seedling growth and some biochemical parameters of dry seeds of black gram. *Int. J. Pharm. Phytopharmacol. Res, 1*, 194-202.

Pratap, A., Chaturvedi, S.K., Tomar, R., Rajan, N., Malviya, N., Thudi, M., Saabale, P.R., Prajapati, U., Varshney, R.K., Singh, N.P. (2017). Marker-assisted introgression of resistance to fusarium wilt race 2 in Pusa 256, an elite cultivar of desi chickpea. *Mol. Genet. Genomics, 292*(6), 1237-1245.
[http://dx.doi.org/10.1007/s00438-017-1343-z] [PMID: 28668975]

Pratap, A., Gupta, S., Malviya, N., Tomar, R., Maurya, R., John, K.J. (2015). Genome scanning of Asiatic Vigna species for discerning population genetic structure based on microsatellite variation. *Mol. Breed., 35*(9), 178.
[http://dx.doi.org/10.1007/s11032-015-0355-9]

Pratap, A., Gupta, S., Tomar, R., Malviya, N., Maurya, R., Pandey, V.R. (2016). Cross-genera amplification of informative microsatellite markers from common bean and scarlet runner bean for assessment of genetic diversity in mungbean (*Vigna radiata*). *Plant Breed., 135*(4), 499-505.
[http://dx.doi.org/10.1111/pbr.12376]

Priya, M., Sharma, L., Kaur, R., Bindumadhava, H., Nair, R.M., Siddique, K.H.M., Nayyar, H. (2019). GABA (γ-aminobutyric acid), as a thermo-protectant, to improve the reproductive function of heat-stressed mungbean plants. *Sci. Rep., 9*(1), 7788.
[http://dx.doi.org/10.1038/s41598-019-44163-w] [PMID: 31127130]

Priya, M., Sharma, L., Singh, I., Bains, T.S., Siddique, K.H.M., H, B., Nair, R.M., Nayyar, H. (2019). Securing reproductive function in mungbean grown under high temperature environment with exogenous application of proline. *Plant Physiol. Biochem., 140*, 136-150.
[http://dx.doi.org/10.1016/j.plaphy.2019.05.009] [PMID: 31103796]

Rainey, K.M., Griffiths, P.D. (2005). Inheritance of heat tolerance during reproductive development in snap

bean (*Phaseolus vulgaris L.*). *J. Am. Soc. Hortic. Sci., 130*(5), 700-706.
[http://dx.doi.org/10.21273/JASHS.130.5.700]

Rasheed, A., Wen, W., Gao, F., Zhai, S., Jin, H., Liu, J., Guo, Q., Zhang, Y., Dreisigacker, S., Xia, X., He, Z. (2016). Development and validation of KASP assays for genes underpinning key economic traits in bread wheat. *Theor. Appl. Genet., 129*(10), 1843-1860.
[http://dx.doi.org/10.1007/s00122-016-2743-x] [PMID: 27306516]

Rasheed, R., Wahid, A., Farooq, M., Hussain, I., Basra, S.M. (2011). Role of proline and glycinebetaine pretreatments in improving heat tolerance of sprouting sugarcane (Saccharum sp.) buds. *Plant Growth Regul., 65*(1), 35-45.
[http://dx.doi.org/10.1007/s10725-011-9572-3]

Rawson, H.M., Craven, C.L. (1979). Variation between short duration mungbean cultivars (*Vigna radiata (L.) Wilczek*) in response to temperature and photoperiod. *Indian J. Plant. Physiol., 22*, 127-136.

Ray, D., Sheshshayee, M.S., Mukhopadhyay, K., Bindumadhava, H., Prasad, T.G., Kumar, M.U. (2003). High nitrogen use efficiency in rice genotypes is associated with higher net photosynthetic rate at lower Rubisco content. *Biol. Plant., 46*(2), 251-256.
[http://dx.doi.org/10.1023/A:1022858828972]

Reardon, M.E., Qaderi, M.M. (2017). Individual and interactive effects of temperature, carbon dioxide and abscisic acid on mung bean (*Vigna radiata*) plants. *J. Plant Interact., 12*(1), 295-303.
[http://dx.doi.org/10.1080/17429145.2017.1353654]

Reddy, K.S. (2009). A new mutant for yellow mosaic virus resistance in Mungbean (Vigna radiata (L.) Wilczek) variety SML-668 by Recurrent Gamma-ray Irradiation *Induced Plant Mutations in the Genomics Era* (pp. 361-362). Rome: Food and Agriculture Organization of the United Nation.

Ruan, Y.L., Jin, Y., Yang, Y.J., Li, G.J., Boyer, J.S. (2010). Sugar input, metabolism, and signaling mediated by invertase: roles in development, yield potential, and response to drought and heat. *Mol. Plant, 3*(6), 942-955.
[http://dx.doi.org/10.1093/mp/ssq044] [PMID: 20729475]

Sakata, T., Oshino, T., Miura, S., Tomabechi, M., Tsunaga, Y., Higashitani, N., Miyazawa, Y., Takahashi, H., Watanabe, M., Higashitani, A. (2010). Auxins reverse plant male sterility caused by high temperatures. *Proc. Natl. Acad. Sci. USA, 107*(19), 8569-8574.
[http://dx.doi.org/10.1073/pnas.1000869107] [PMID: 20421476]

Saleh, A.A., Abdel-Kader, D.Z., El Elish, A.M. (2007). Role of heat shock and salicylic acid in antioxidant homeostasis in mungbean (*Vigna radiata L.*) plant subjected to heat stress. *Am. J. Plant Physiol., 2*(6), 344-355.
[http://dx.doi.org/10.3923/ajpp.2007.344.355]

Schafleitner, R., Nair, R.M., Rathore, A., Wang, Y.W., Lin, C.Y., Chu, S.H., Lin, P.Y., Chang, J.C., Ebert, A.W. (2015). The AVRDC - The World Vegetable Center mungbean (*Vigna radiata*) core and mini core collections. *BMC Genomics, 16*(1), 344.
[http://dx.doi.org/10.1186/s12864-015-1556-7] [PMID: 25925106]

Sehrawat, N., Jaiwal, P.K., Yadav, M., Bhat, K.V., Sairam, R.K. (2013). Salinity stress restraining mungbean (*Vigna radiata L.*) Wilczek) production: gateway for genetic improvement. *Int. J. Agric. Crop Sci., 6*(9), 505.

Sgobba, A., Paradiso, A., Dipierro, S., De Gara, L., de Pinto, M.C. (2015). Changes in antioxidants are critical in determining cell responses to short- and long-term heat stress. *Physiol. Plant., 153*(1), 68-78.
[http://dx.doi.org/10.1111/ppl.12220] [PMID: 24796393]

Sharkey, T.D. (2005). Effects of moderate heat stress on photosynthesis: importance of thylakoid reactions, rubisco deactivation, reactive oxygen species, and thermotolerance provided by isoprene. *Plant Cell Environ., 28*(3), 269-277.
[http://dx.doi.org/10.1111/j.1365-3040.2005.01324.x]

Sharma, L., Priya, M., Bindumadhava, H., Nair, R.M., Nayyar, H. (2016). Influence of high temperature

stress on growth, phenology and yield performance of mungbean [*Vigna radiata (L.)* Wilczek] under managed growth conditions. *Sci. Hortic. (Amsterdam), 213*, 379-391.
[http://dx.doi.org/10.1016/j.scienta.2016.10.033]

Siczek, A., Lipiec, J. (2011). Soybean nodulation and nitrogen fixation in response to soil compaction and surface straw mulching. *Soil Tillage Res., 114*(1), 50-56.
[http://dx.doi.org/10.1016/j.still.2011.04.001]

Singh, B., Singh, N., Thakur, S., Kaur, A. (2017). Ultrasound assisted extraction of polyphenols and their distribution in whole mung bean, hull and cotyledon. *J. Food Sci. Technol., 54*(4), 921-932.
[http://dx.doi.org/10.1007/s13197-016-2356-z] [PMID: 28303043]

Singh, D.P., Singh, B.B. (2011). Breeding for tolerance to abiotic stresses in mungbean. *Indian J. Pulses Res., 24*(2), 83-90.
[http://dx.doi.org/10.15373/2249555X/June2014/25]

Sinsawat, V., Leipner, J., Stamp, P., Fracheboud, Y. (2004). Effect of heat stress on the photosynthetic apparatus in maize (Zea mays L.) grown at control or high temperature. *Environ. Exp. Bot., 52*(2), 123-129.
[http://dx.doi.org/10.1016/j.envexpbot.2004.01.010]

Sita, K., Sehgal, A., Hanumantha Rao, B., Nair, R. M., Vara Prasad, P. V., Kumar, S., Nayyar, H. (2017). Food legumes and rising temperatures: effects, adaptive functional mechanisms specific to reproductive growth stage and strategies to improve heat tolerance. *Front. Plant Sci.,* 1658-8.

Sita, K., Sehgal, A., Kumar, J., Kumar, S., Singh, S., Siddique, K.H.M., Nayyar, H. (2017). Identification of high-temperature tolerant lentil (Lens culinaris Medik.) genotypes through leaf and pollen traits. *Front. Plant Sci., 8*, 744.
[http://dx.doi.org/10.3389/fpls.2017.00744] [PMID: 28579994]

Srinivasan, A., Takeda, H., Senboku, T. (1996). Heat tolerance in food legumes as evaluated by cell membrane thermostability and chlorophyll fluorescence techniques. *Euphytica, 88*(1), 35-45.
[http://dx.doi.org/10.1007/BF00029263]

Taiz, L., Zeiger, E. (2006). Fisiologia vegetal. *Universitat Jaume I., Vol. 10*, 1265.

Team, C.W., Pachauri, R.K., Meyer, L.A. (2014). *IPCC, 2014: climate change 2014: synthesis report. Contribution of Working Groups I. II III to Fifth Assess Rep Intergov panel Clim Chang IPCC.* (p. 151). Geneva: Switz.

Teixeira, E.I., Fischer, G., Van Velthuizen, H., Walter, C., Ewert, F. (2013). Global hot-spots of heat stress on agricultural crops due to climate change. *Agric. For. Meteorol., 170*, 206-215.
[http://dx.doi.org/10.1016/j.agrformet.2011.09.002]

Thirumaran, A. S., Seralathan, M. A. *Utilization of mungbean (No. RESEARCH).*AVRDC. (1988).

Vadez, V., Berger, J.D., Warkentin, T., Asseng, S., Ratnakumar, P., Rao, K.P.C. (2012). Adaptation of grain legumes to climate change: a review. *Agron. Sustain. Dev., 32*(1), 31-44.
[http://dx.doi.org/10.1007/s13593-011-0020-6]

Van Breusegem, F., Vranová, E., Dat, J.F., Inzé, D. (2001). The role of active oxygen species in plant signal transduction. *Plant Sci., 161*(3), 405-414.
[http://dx.doi.org/10.1016/S0168-9452(01)00452-6] [PMID: 11166426]

Varshney, R.K., Terauchi, R., McCouch, S.R. (2014). Harvesting the promising fruits of genomics: applying genome sequencing technologies to crop breeding. *PLoS Biol., 12*(6)e1001883
[http://dx.doi.org/10.1371/journal.pbio.1001883] [PMID: 24914810]

Vinocur, B., Altman, A. (2005). Recent advances in engineering plant tolerance to abiotic stress: achievements and limitations. *Curr. Opin. Biotechnol., 16*(2), 123-132.
[http://dx.doi.org/10.1016/j.copbio.2005.02.001] [PMID: 15831376]

Wang, H., Wang, H., Shao, H., Tang, X. (2016). Recent advances in utilizing transcription factors to improve plant abiotic stress tolerance by transgenic technology. *Front. Plant Sci., 7*, 67.

[http://dx.doi.org/10.3389/fpls.2016.00067] [PMID: 26904044]

Wang, X., Dinler, B.S., Vignjevic, M., Jacobsen, S., Wollenweber, B. (2015). Physiological and proteome studies of responses to heat stress during grain filling in contrasting wheat cultivars. *Plant Sci., 230*, 33-50. [http://dx.doi.org/10.1016/j.plantsci.2014.10.009] [PMID: 25480006]

Waraich, E.A., Ahmad, R., Halim, A., Aziz, T. (2012). Alleviation of temperature stress by nutrient management in crop plants: a review. *J. Soil Sci. Plant Nutr., 12*(2), 221-244. [http://dx.doi.org/10.4067/S0718-95162012000200003]

Yadav, S.S., McNeil, D.L., Redden, R., Patil, S.A. (2010). *Climate Change and Management of Cool Season Grain Legume Crops.* Springer Science & Business Media. [http://dx.doi.org/10.1007/978-90-481-3709-1]

Young, L.W., Wilen, R.W., Bonham-Smith, P.C. (2004). High temperature stress of Brassica napus during flowering reduces micro- and megagametophyte fertility, induces fruit abortion, and disrupts seed production. *J. Exp. Bot., 55*(396), 485-495. [http://dx.doi.org/10.1093/jxb/erh038] [PMID: 14739270]

Zahran, H.H. (1999). Rhizobium-legume symbiosis and nitrogen fixation under severe conditions and in an arid climate. *Microbiol. Mol. Biol. Rev., 63*(4), 968-989. [http://dx.doi.org/10.1128/MMBR.63.4.968-989.1999] [PMID: 10585971]

Zandalinas, S.I., Mittler, R., Balfagón, D., Arbona, V., Gómez-Cadenas, A. (2018). Plant adaptations to the combination of drought and high temperatures. *Physiol. Plant., 162*(1), 2-12. [http://dx.doi.org/10.1111/ppl.12540] [PMID: 28042678]

Zheng, B., Chenu, K., Fernanda Dreccer, M., Chapman, S.C. (2012). Breeding for the future: what are the potential impacts of future frost and heat events on sowing and flowering time requirements for Australian bread wheat (Triticum aestivium) varieties? *Glob. Change Biol., 18*(9), 2899-2914. [http://dx.doi.org/10.1111/j.1365-2486.2012.02724.x] [PMID: 24501066]

Zinn, K.E., Tunc-Ozdemir, M., Harper, J.F. (2010). Temperature stress and plant sexual reproduction: uncovering the weakest links. *J. Exp. Bot., 61*(7), 1959-1968. [http://dx.doi.org/10.1093/jxb/erq053] [PMID: 20351019]

# SUBJECT INDEX

# L

Laboratory-cum-field screening technique 77
Large genetic variability for 80, 97
    heat tolerance 97
    tolerance 80
Late embryogenesis abundant (LEA) 5, 102
Leaf 7, 43, 96
    area index (LAI) 7, 96
    firing phenotypes 43

# M

Maize 30, 32, 33, 34, 40, 42, 43, 44, 49, 50, 52
    effects of heat stress on 30, 32
    heat stress in 32, 33, 34, 40, 49
    heat-tolerant 43, 44
    metabolite profiling of 40
    screening of 42, 52
    spring 31, 43, 44
    transgenic 50
Maize breeding 45
    programme, global 45
Maize 28, 32, 38, 39, 41, 49, 52
    chloroplasts 38
    epigenome 39
    genes 49
    grain 32, 52
    growth 28, 41
    HSP genes 49
    transgenics lines 49
Maize cultivars 28, 53
    breeding heat-tolerant 28
Maize genotypes 28, 34, 37, 40, 41, 44, 45, 50, 51, 52, 53
    heat-tolerant 45
    heat-tolerant sweet 40
Maize germplasm 44, 52, 53
    heat-tolerant 44
Maize hybrids 34
    heat resilient 28
    heat stress tolerant 50
    heat-tolerant 30, 45, 51, 52, 53
    stress-tolerant 45
Marker-assisted recurrent selection (MARS) 80
Megasporogenesis 94
Membrane-associated thermosensors 36

Membrane 72, 36
    fluidity 72
    fluidizer benzyl alcohol 36
Membrane integrity 98, 122, 153, 158
    affected 153
    cellular 122
Membrane lipid 4, 37, 120, 121
    composition 37, 120, 121
    fractions 4
Membrane stability 9, 32, 37, 46, 75, 76, 100, 131, 134
    cellular 46
    decreased cell 76
    plasma 131
    index (MSI) 100
Membrane thermostability 43, 122
    tests 122
    traits 43
Metabolites 2, 41, 68, 74, 75, 97, 123, 125, 128, 134, 145
    toxic 145
    tryptophan 41
Method 2, 6, 78, 96, 144
    acetone extraction 6
    agronomic management 144
    automated sequencing 2
    chlorophyll florescence 78
    embryo protein synthesis 78
    gene characterization 144
    non-destructive 6, 96
Microsporogenesis 94
Modification 14, 50, 121, 159
    chromatin 14
    genetic 159
    histone 14
Molecular chaperones 12, 35, 38, 126
Molecules 1, 2, 5, 13, 35, 74, 121, 125, 145, 160
    cellular 2
    identification polar 13
    osmoprotectant 1, 5
    protective 160, 161
    signaling 121, 125, 160
MTS and chlorophyll 78, 99
    content 99
    fluorescence 78
Mungbean 148, 151, 153, 158, 159, 162
    cultivars 148, 153, 158
    genome 162
    germplasm 159

# V

# W

# Z